普通高等教育"十三五"规划教材

JSP 程序设计项目教程

主　编　王平华　徐卫红　邹金萍
副主编　刘志华　李雅萍　曾鹏程
　　　　谢忠瑞　傅　勇　周　峰
主　审　周学军

电子工业出版社
Publishing House of Electronics Industry
北京·BEIJING

内 容 简 介

本书以帮助读者熟练掌握 JSP 技术的具体应用为目标。全书共 11 章,提供了多个真实的项目案例,包括手机信息采集、验证码、用户身份验证、剪刀石头布、发牌游戏、JavaBean 与动作指令应用、JSON 与 JavaBean 转换应用、基于 Servlet+MD5+盐值的用户登录、基于 MVC 的三层架构用户管理、日志文件管理、单点登录及授权访问、基于单例模式及缓冲池的 JDBC 数据库开发技术、多文件上传技术、安全的文件下载技术等。每个案例从【案例描述】→【案例分析】→【案例实现】→【运行结果】几个方面进行了全面剖析,内容由浅入深,让读者在项目实战中成长。

本书附有案例、学习网站、教学课件、实训开发框架等教学资源,而且为了帮助初学者更好地学习本书讲解的内容,还提供了在线答疑,希望可以帮助更多读者,详见前言。

本书不仅可以作为高职高专院校计算机类专业的教材,也可以作为软件开发人员和计算机爱好者的参考用书。

未经许可,不得以任何方式复制或抄袭本书之部分或全部内容。
版权所有,侵权必究。

图书在版编目(CIP)数据

JSP 程序设计项目教程 / 王平华,徐卫红,邹金萍主编. —北京:电子工业出版社,2019.9

ISBN 978-7-121-36490-7

Ⅰ. ①J… Ⅱ. ①王… ②徐… ③邹… Ⅲ. ①JAVA 语言—网页制作工具—程序设计—高等职业教育—教材
Ⅳ. ①TP312②TP393.092

中国版本图书馆 CIP 数据核字(2019)第 089229 号

责任编辑:胡辛征
印　　刷:天津画中画印刷有限公司
装　　订:天津画中画印刷有限公司
出版发行:电子工业出版社
　　　　　北京市海淀区万寿路 173 信箱　邮编　100036
开　　本:787×1 092　1/16　印张:17.5　字数:448 千字
版　　次:2019 年 9 月第 1 版
印　　次:2021 年 8 月第 4 次印刷
定　　价:49.80 元

凡所购买电子工业出版社图书有缺损问题,请向购买书店调换。若书店售缺,请与本社发行部联系,联系及邮购电话:(010)88254888,88258888。
质量投诉请发邮件至 zlts@phei.com.cn,盗版侵权举报请发邮件至 dbqq@phei.com.cn。
本书咨询联系方式:peijie@phei.com.cn。

前 言

"JSP 程序设计"是软件技术专业的核心课程,本书是为学习 JSP 程序设计课程而设计的项目化教材,该教材以培养学生的 Java Web 项目开发能力为导向,通过实际项目案例阐述了 JSP 的工作原理、JSP 编程技术、JSP 动作指令、内置对象技术、EL 与 JSTL 标签、JavaBean 技术、Servlet 技术、过滤器与监听器、JDBC 数据库开发、文件上传与下载等技术的使用规范及项目应用。它与目前在售的同类型图书的不同之处在于:一是本书项目案例技术均来自于企业;二是本书注重项目开发能力的培养,它不是技术语法及 API 的参考书;三是本书提供了一套用于 JSP 实训或 Java Web 项目开发的软件开发框架。

本书共包含多个企业级项目案例,最大的特色是"技术实用、易教易学"。

1. 真实的项目案例

本书作者根据十多年 Java 项目开发经验作为教材编写背景,采用真实的项目案例引导读者理解 JSP 技术的具体应用,本书不仅注重知识的传递,更强调项目开发能力的培养,全书提供了手机信息采集、验证码、用户身份验证、剪刀石头布、发牌游戏、JavaBean 与动作指令应用、JSON 与 JavaBean 转换应用、基于 Servlet+MD5+盐值的用户登录、基于 MVC 的三层架构用户管理、日志文件管理、单点登录及授权访问、基于单例模式及缓冲池的 JDBC 数据库开发技术、多文件上传技术、安全的文件下载技术项目案例,每个案例均独立,由浅入深、实例生动、易学易用,让读者在项目实战中成长。

2. 搭建完善的项目开发框架

为了方便教师和学生进行 JSP 程序设计课程实训,本书提供了一套适合 Java Web 项目开发的框架,框架不仅整合了本书中的所有技术,而且将一般 Java Web 项目开发所需的技术进行封装,利用它可以快速进行实战项目开发。

3. 通过纸质教材、课程学习网站、教学课件共同打造立体化教材

丰富的网站资源和图文并茂的教学课件为"教"和"学"提供了最大便利,为了帮助读者理解项目案例的设计思路,作者提供了许多原创图片,并配以文字辅助,以便读者能深入分析并理解问题;项目案例、课程网站、教学课件、实训开发框架等形成了一套立体化的教学资源。以上教学资源,读者均可访问华信教育资源网(www.hxedu.com.cn)免费下载使用。

本书是一本校企合作、工学结合的教材,由王平华、徐卫红、邹金萍担任主编,刘志华、李雅萍、曾鹏程、谢忠瑞、傅勇、周峰担任副主编,周学军担任主审。作者王平华系国内高职院校的一线专业教师,兼任软件技术公司技术总监,编写教材时得到了思创数码科技股份有限公司及江西联微软件技术有限公司的大力支持。具体分工如下:李雅萍编写

第 1、2 章，刘志华编写第 3、4 章，徐卫红编写第 5、6 章，王平华编写第 7~11 章，邹金萍、曾鹏程、谢忠瑞、傅勇、周峰参与了本书部分内容的编写，本书由刘伟杰审稿，在此一并表示感谢！

  由于编者水平有限，书中难免存在错误或不妥之处，恳请广大读者提出宝贵意见，作者的 E-mail：software_book@163.com。同时欢迎加入软件技术专业教师 QQ 交流群（群号：528948207），群内提供教学所需全部资料，并将持续提供教学与技术支持。

编 者

目 录

第 1 章 Web 应用程序开发概述 ... 1
1.1 网络应用程序结构的演变 ... 1
1.1.1 B/S 结构和 C/S 结构 ... 1
1.1.2 Web 应用程序 ... 3
1.1.3 Web 的有关概念 ... 4
1.2 网页的类型和工作原理 ... 7
1.2.1 静态网页和动态网页 ... 7
1.2.2 为什么需要动态网页 ... 9
习题 ... 9

第 2 章 JSP 工作原理、开发环境及运行配置 ... 10
2.1 JSP 技术概述 ... 10
2.1.1 Java 语言 ... 10
2.1.2 Servlet 技术 ... 11
2.1.3 JavaBean 技术 ... 11
2.1.4 JSP 技术 ... 12
2.1.5 JSP 和 Java Servlet 的关系 ... 12
2.1.6 JSP 在 JavaWeb 开发中的地位 ... 13
2.2 JSP 工作原理 ... 13
2.3 JSP 开发环境搭建 ... 14
2.3.1 JSP 的运行环境 ... 14
2.3.2 JDK 的安装与配置 ... 15
2.3.3 Tomcat 7 的安装与配置 ... 17
2.4 JSP 开发工具 ... 21
2.4.1 IDEA 简介 ... 21
2.4.2 IDEA 的安装及配置 ... 21
2.5 创建第一个 JSP 应用 ... 23
习题 ... 29

第 3 章　JSP 编程基础 ·· 30

3.1　JSP 编程语法 ·· 30
3.1.1　JSP 页面的基本结构 ·································· 30
3.1.2　JSP 变量的声明 ·· 32
3.1.3　选择语句 ·· 33
3.1.4　循环语句 ·· 33

3.2　JSP 页面编程 ·· 34
3.2.1　脚本程序 ·· 34
3.2.2　变量与方法的声明 ···································· 35
3.2.3　JSP 表达式 ··· 35
3.2.4　JSP 中的注释 ·· 35

3.3　输出 26 个英文字母 ·· 36
3.4　抽奖游戏 ·· 37
习题 ·· 39

第 4 章　JSP 指令操作 ·· 40

4.1　编译指令 ·· 40
4.1.1　page 指令 ·· 41
4.1.2　include 指令 ··· 43
4.1.3　taglib 指令 ·· 44

4.2　动作指令 ·· 45
4.2.1　include 指令 ·· 45
4.2.2　useBean 指令 ·· 45
4.2.3　setPoperty 指令 ·· 46
4.2.4　getPoperty 指令 ·· 47
4.2.5　forward 指令 ··· 47
4.2.6　plugin 指令 ··· 47

4.3　设计一个登录页面 ·· 48
4.4　Excel 解析收到的信息 ···································· 52
习题 ·· 53

第 5 章　内置对象技术 ·· 54

5.1　内置对象概述 ·· 54
5.1.1　request 对象 ·· 55

	5.1.2	out 内置对象	56
	5.1.3	response 内置对象	57
	5.1.4	session 内置对象	59
	5.1.5	application 内置对象	60
	5.1.6	page 内置对象	62
	5.1.7	pageContext 内置对象	62
	5.1.8	config 内置对象	62
	5.1.9	exception 内置对象	62

5.2 内置对象的使用 62
    5.2.1 手机信息采集 63
    5.2.2 验证码 75
    5.2.3 用户身份验证 79
习题 86

第 6 章 EL 与 JSTL 标签 87

6.1 EL 与 JSTL 概述 87
    6.1.1 EL 概述 88
    6.1.2 JSTL 概述 89

6.2 剪刀石头布游戏 95
6.3 发牌游戏 99
习题 107

第 7 章 JavaBean 技术 108

7.1 JavaBean 概述 108
    7.1.1 JavaBean 组成 109
    7.1.2 JavaBean 作用范围 109

7.2 JavaBean 与动作指令应用 110
7.3 JSON 与 JavaBean 转换应用 114
习题 120

第 8 章 Servlet 技术 121

8.1 Servlet 相关知识 121
    8.1.1 Servlet 相关类 122
    8.1.2 Servlet 类定义方式 123

8.2 基于 Servlet 用户登录 125

8.3 基于 MVC 的三层架构用户管理 ··129
习题 ··154

第 9 章 过滤器与监听器 ··156

9.1 过滤器与监听器相关知识 ···156
  9.1.1 过滤器 ···157
  9.1.2 监听器 ···159
9.2 日志文件 ···161
9.3 单点登录及授权访问 ··168
习题 ··180

第 10 章 JDBC 数据库开发 ··182

10.1 JDBC 相关知识 ··182
  10.1.1 JDBC 核心类 ···183
  10.1.2 JDBC 连接池配置 ···187
  10.1.3 单例模式 DBHelper 类 ···188
10.2 用户 CRUD 开发 ···193
10.3 登录与 MD5 密码管理 ···221
习题 ··226

第 11 章 文件上传与下载 ··227

11.1 文件上传与下载相关知识 ··227
  11.1.1 文件上传相关知识 ···228
  11.1.2 文件下载相关知识 ···230
11.2 文件上传 ··238
11.3 文件下载 ··265
习题 ··271

第 1 章

Web 应用程序开发概述

随着互联网技术的应用和普及,各行各业对开发 Web 应用程序的需求高涨。简单地说,Web 应用程序是一种基于 B/S 结构的网络软件,它使用 HTTP 协议作为通信协议,通过网络让浏览器与服务器进行通信的计算机程序。

本章任务

(1) 网络应用程序结构的演变;
(2) 网页的类型和工作原理。

重点内容

(1) B/S 结构与 C/S 结构;
(2) Web 应用程序。

难点内容

静态网页和动态网页的区别。

1.1 网络应用程序结构的演变

1.1.1 B/S 结构和 C/S 结构

早期的应用程序都是运行在单机上的,称为桌面应用程序。后来由于网络的普及,出现了运行在网络上的网络应用程序(网络软件),网络应用程序有 C/S 和 B/S 两种结构。

1. C/S 结构

C/S 是 Client/Server 的缩写,C/S 结构即客户机/服务器结构,这种结构的软件包括客户端程序和服务器端程序两部分。就像人们常用的 QQ 等网络软件，需要下载并安装专用的客户端软件，并且服务器端也需要特定的软件支持才能运行，QQ 客户端页面如图 1-1 所示。

图 1-1 QQ 客户端页面

C/S 结构最大的缺点是不易于部署。因为每台客户端都要安装客户端软件。如果客户端软件需要升级，则必须为每台客户端单独升级。另外，客户端软件通常对客户机的操作系统也有要求，如有些客户端软件只能运行在 Windows 平台下。

2．B/S 结构

B/S 是 Browser/Server 的缩写，B/S 结构即浏览器/服务器结构。它是随着 Internet 技术的兴起，对 C/S 结构的一种变化或改进的结构。在这种结构下，客户端软件由浏览器来代替，B/S 结构的浏览器端页面如图 1-2 所示，一部分事务逻辑在浏览器端（Browser）实现，但是主要事务逻辑在服务器端（Server）实现。目前流行的是三层 B/S 结构，即表现层、事务逻辑层和数据处理层。

图 1-2 B/S 结构的浏览器端页面

B/S 结构很好地解决了 C/S 结构的上述缺点。因为每台客户端都自带浏览器，所以就不需要额外安装客户端软件了，也就不存在客户端软件升级的问题了。另外，由于任何操作系统一般都带有浏览器，因此 B/S 结构对客户端的操作系统也没有特殊的要求。

但是 B/S 结构与 C/S 结构相比，也有其自身的缺点。首先，因为 B/S 结构的客户端 8F6F 件页面就是网页，因此操作页面不可能做得很复杂、漂亮。例如，很难实现树形菜单、选项卡式面板或右键快捷菜单等（或者虽然能够模拟实现，但是响应速度比 C/S 结构中的客户端软件要慢很多）；其次，B/S 结构下的每次操作一般都要刷新网页，响应速度明显不如 C/S 结构；再次，在网页操作页面下，大多以鼠标操作方式为主，无法定义快捷键，也就无法满足快速操作的需求。

> 提示：因为 C/S 结构和 B/S 结构的网络软件其程序都分布在客户机和服务器上，因此它们统称为分布式系统（Distributed System）。

1.1.2 Web 应用程序

Web 应用程序是 B/S 结构的软件产物，它首先是"应用程序"，和 C、C++等程序设计语言编写出来的程序没有本质的区别。然而 Web 应用程序又有其自身独特的地方，主要表现在：

① Web 应用程序是基于 Web 的，依赖于通用的 Web 浏览器来表现它的执行结果；
② 需要一台 Web 服务器，在服务器上对数据进行处理，并将处理结果生成网页，以方便客户端直接使用浏览器浏览。

1. Web 应用程序与网站

一般来说，网站的内容需要经常更新，并添加新内容。早期的网站是静态的，更新静态网站的内容是非常烦琐的。例如，要增加一个新网页，就需要手工编辑这个网页的 HTML 代码，然后更新相关页面到这个页面的链接，最后把所有更新过的页面重新上传到服务器上。

为了提高网站内容更新的效率，就需要构建 Web 应用程序来管理网站内容。Web 应用程序可以把网站的 HTML 页面部分和数据部分分离开。要更新或添加新网页，只要在数据库中更新或添加记录就可以了，程序会自动读取数据库中的记录，生成新的页面代码发送给浏览器，从而实现了网站内容的动态更新。

可见，Web 应用程序能够动态生成网页代码，Web 应用程序可以通过各种服务器端脚本语言来编写。而服务器端脚本代码是可以嵌入网页的 HTML 代码的，嵌入了服务器端脚本代码的网页就称为动态网页文件。因此，如果一个网站中含有动态网页文件，则这个网站就相当于是一个 Web 应用程序。

利用 Web 应用程序，网站可以实现动态更新页面，以及与用户进行交互（如留言板、论坛博客、发表评论）等各种功能。但 Web 应用程序并不等同于动态网站，它们的侧重点不同。一般来说，动态网站侧重于给用户提供信息，而 Web 应用程序侧重于完成某种特定任务，如基于 B/S 结构的管理信息系统（Management Information System，MIS）就是一种 Web 应用程序，但不能称为网站。Web 应用程序的真正核心功能是对数据库进行处理。

2. Web 应用程序的组成

Web 应用程序通常由 HTML 文件、服务器端脚本文件和一些资源文件组成。

HTML 文件可以提供静态的网页内容。服务器端脚本文件可以提供程序，实现客户端与服务器之间的交互，以及访问数据库。资源文件可以是图片文件、多媒体文件和配置文件等。

3. 运行 Web 应用程序的要素

要运行 Web 应用程序，需要考虑 Web 服务器、浏览器和 HTTP 通信协议等因素。

（1）Web 服务器。

运行 Web 应用程序需要一个载体，即 Web 服务器。一个 Web 服务器可以放置多个 Web 应用程序。

通常 Web 服务器有两层含义：一方面它代表运行 Web 应用程序的计算机硬件设备，一台计算机只要安装了操作系统和 Web 服务器软件，就可算作一台 Web 服务器；另一方面，Web 服务器专指一种软件——Web 服务器软件，该软件的功能是响应用户通过浏览器提交的 HTTP 请求。如果用户请求的是 JSP 脚本，则 Web 服务器软件将解析并执行 JSP 脚本，生成 HTML 格式的文本，并发送到客户端，显示在浏览器中。

（2）浏览器。

浏览器是用于解析 HTML 文件（可包括 CSS 代码和客户端 JavaScript 脚本）并显示的应用程序，它可以从 Web 服务器接收、解析和显示信息资源（可以是网页或图像等），信息资源一般使用统一资源定位符（URL）标识。

浏览器只能解析和显示 HTML 文件，而无法处理服务器端脚本文件（如 JSP 文件），这就是为什么可以直接用浏览器打开 HTML 网页文件，而服务器端脚本文件只有被放置在 Web 服务器上，才能被正常浏览的原因。

（3）HTTP 通信协议（超文本传输协议）。

HTTP 是浏览器与 Web 服务器之间通信的语言。浏览器与服务器之间的会话（图 1-3），总是由浏览器向服务器发送 HTTP 请求信息开始的（如用户输入网址，请求某个网页文件），Web 服务器根据请求返回相应的信息，这称为 HTTP 响应，响应中包含请求的完整状态信息，并在消息体中包含请求的内容（如用户请求的网页文件内容等）。

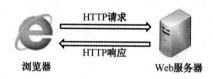

图 1-3　浏览器与服务器之间的对话

1.1.3　Web 的有关概念

在学习 Web 编程前，有必要明确 URL、域名、HTTP 和 MIME 这些概念。

1. URL

当用户使用浏览器访问网站时，通常都会在浏览器的地址栏中输入网站地址，这个地址就是 URL（Universal Resource Locator，统一资源定位符）。URL 信息会通过 HTTP 请求发送给服务器，服务器根据 URL 信息返回对应的网页文件代码给浏览器。

URL 是 Internet 上任何资源的标准地址，每个网站上的每个网页（或其他文件）在 Internet 上都有唯一的 URL 地址，通过网页的 URL，浏览器就能定位到目标网页或资源文件。

URL 的一般格式为："协议名://主机名[:端口号][/目录路径/文件名][#锚点名]"

图 1-4 是一个 URL 的结构示例。

图 1-4　URL 的结构示例

URL 后必须接":∥",其他各项之间用"/"隔开。例如,图 1-4 中的 URL 表示信息被放在一台被称为 www 的服务器上,hynu.cn 是一个已被注册的域名,.cn 表示中国,主机名、域名合称为主机头;web/201009/是服务器网站目录下的目录路径,而 first.html 是位于上述目录下的文件名,因此该 URL 能够让用户访问到这个文件。

在 URL 中,常见的"协议"有 HTTP 和 FTP。

(1) HTTP:超文本传输协议,用于传送网页。例如:

http://bbs.runsky.com:8080/bbs/display.jsp#fid

(2) FTP:文件传输协议,用于传送文件。例如:

ftp://219.216.128.15/
ftp://001.seaweb.cn/web

2. 域名

在 URL 中,主机名通常是域名或 IP 地址。最初,域名是为了方便人们记忆 IP 地址的,使用户在 URL 中可以输入域名而不必输入难记的 IP 地址。但现在多个域名可对应一个 IP 地址(一台主机),即在一台主机上可架设多个网站,这些网站的存放方式称为"虚拟主机"方式,此时由于一个 IP 地址(一台主机)对应多个网站,就不能采用输入 IP 地址的方式访问网站,而只能在 URL 中输入域名。Web 服务器为了区别用户请求的是这台主机上的哪个网站,通常必须为每个网站设置"主机头"来区别这些网站。

因此域名的作用有两个:一是将域名发送给 DNS 服务器解析得到域名对应的 IP 地址以进行连接;二是将域名信息发送给 Web 服务器,通过域名与 Web 服务器上设置的"主机头"进行匹配确认客户端请求的是哪个网站。浏览器输入网址访问网站的过程如图 1-5 所示。若客户端没有发送域名信息给 Web 服务器,如直接输入 IP,则 Web 服务器将打开服务器上的默认网站。

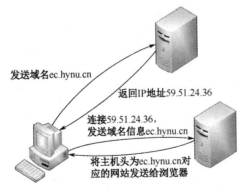

图 1-5　浏览器输入网址访问网站的过程

3. HTTP

HTTP 是用于从浏览器请求 Web 服务器,Web 服务器再传输超文本(或其他文档)到浏览器的传输协议。它不仅能保证计算机正确快速地传输网页文档，还能确定传输文档中的哪一部分，以及哪部分内容首先显示（如文本先于图形）等。这就是在浏览器中看到的网页地址都是以"http://"开头的原因。

HTTP 包含两个阶段：请求阶段和响应阶段。浏览器和 Web 服务器之间的每次 HTTP 通信（请求或响应），都包含两部分：头部和主体。头部包含了与通信有关的信息；主体则包含通信的数据，当然，前提是存在这样的数据。

（1）HTTP 请求阶段。

HTTP 请求的通用格式如下：

①首行:HTTP方法　　　URL中的域名部分　　　HTTP版本
②头部字段
③空行
④消息主体

下面是一个 HTTP 请求首行的示例：

```
GET / content. html HTTP/1.1
```

它表示使用 GET 方式向服务器请求 content.html 这个文档，使用的协议是 HTTP 1.1 版本。对于 HTTP 方法来说，最常用的是 GET 和 POST 两种方法。GET 方法用来请求服务器返回指定文档的内容；POST 方法表示发送附加的数据并执行指定的文档，它最常见的应用是从浏览器向服务器发送表单数据，同时还发送一个请求执行服务器中的某个程序（动态页），这个程序将处理这些表单数据。

第二部分是头部字段，一个常用的头部请求字段为 Accept 字段，该字段用来指定浏览器可以接受哪些类型的文档。例如，Accept: text/html 表示浏览器只可接受 HTML 文档。文档类型采用 MIME 类型来表示。如果浏览器可以接受多种格式的文档，那么可以指定多个 Accept 字段。

请求的头部字段之后必须有一个空行，该空行用于将请求的主体和头部分隔开来。使用了 GET 方法的请求没有请求主体。因此，这种情况下，空行是请求结束的标记。

（2）HTTP 响应阶段。

HTTP 响应的通用格式如下：

①状态行
②响应头部字段
③空行
④响应主体

状态行中包含了所用 HTTP 的版本号，此外还包括一个用 3 位数表示的响应状态码，以及针对状态码的一个简短的文本解释。例如，大部分响应都是以下面的状态行开头的。

```
HTTP/1.1 200 OK
```

它表示响应使用的协议是 HTTP 1.1。状态码是 200，文本解释是 OK。

其中，状态码 200 表示请求得到处理，没有发生任何错误，这是用户希望看到的。状态码 404 表示请求的文件未找到；状态码 500 表示服务器出现了错误，且不能完成请求。

状态行之后是响应头部字段，响应头部可能包含多行有关响应的信息，每条信息都对应一个字段。响应头部中必须使用的字段只有一个，即 Content-type。例如：

```
Content- type: text/html, charset=UTF-8
```

它表示响应的内容是 HTML 文档，内容采用的编码方式是 UTF-8。

响应头部字段之后必须有一个空行，这与请求头部是一致的。空行之后才是响应主体。在上例中，响应主体是一个 HTML 文件。

4. MIME

浏览器从服务器接收返回的文档时，必须确定这个文档属于哪种格式。如果不了解文档的格式，浏览器将无法正确显示该文档，因为不同的文档格式要求使用不同的解析工具。例如，服务器返回的是一个 JPG 图片格式的文档，而浏览器把它当成 HTML 文档去解析。则显示出来的将是乱码。通过多用途网际邮件扩充协议（MIME）可以指定文档的格式。

MIME 最初的目标是允许各种不同类型的文档都可以通过电子邮件发送。这些文档可能包含各种类型的文本、视频数据或音频数据。由于 Web 服务器也存在这方面的需求，因此 Web 服务器中也采用了 MIME 来指定所传递的文档类型。

Web 服务器在一个将要发送到浏览器的文档头部附加了 MIME 的格式说明。当浏览器从 Web 服务器中接收到这个文档时，就根据其中包含的 MIME 格式说明来确定下一步的操作。例如，如果文档内容为文本，则 MIME 格式说明将通知浏览器文档的内容是文本，并指明具体的文本类型。MIME 说明的格式如下：

类型/子类型

最常见的 MIME 类型为 text（文本）、image（图片）和 video（视频）。其中，最常用的文本子类型为 plain、html 和 xml。最常用的图片子类型为 gif 和 jpeg。服务器通过将文件的扩展名作为类型表中的键值来确定文档的类型。例如，扩展名.html 意味着服务器应该在将文档发送给浏览器之前为文档附加 MIME 说明：text/ html。

1.2 网页的类型和工作原理

1.2.1 静态网页和动态网页

在 Internet 发展初期，Web 上的内容都是由静态网页组成的，Web 开发就是编写一些简单的 HTML 页面，页面上包含一些文本、图片等信息资源，用户可以通过超链接浏览信息。采用静态网页的网站有很明显的局限性，如不能与用户进行交互，不能实时更新网页上的内容。因此像用户留言、发表评论等功能都无法实现，只能做一些简单的展示型网站。

后来 Web 开始由静态网页向动态网页转变，这是 Web 技术经历的一次重大变革。随着动态网页的出现，用户能与网页进行交互，表现在除能浏览网页内容外，还能改变网页内容（如发表评论）。此时用户既是网站内容的消费者（浏览者），又是网站内容的制造者。

1. 静态网页和动态网页的区别

根据 Web 服务器是否需要对网页中脚本代码进行解释（或编译）执行，网页可分为静态网页和动态网页。

（1）静态网页就是纯粹的 HTML 页面，网页的内容是固定的、不变的。用户每次访问静态网页时，其显示的内容都是一样的。

（2）动态网页是指网页中的内容会根据用户请求的不同而发生变化的网页，同一个网页由于每次请求的不同，可显示不同的内容，如图 1-6 和图 1-7 中显示的两个网页实际上是同一个动态网页文件。动态网页中可以变化的内容称为动态内容，它是由 Web 应用程序来实现的。

2. 静态网页的工作流程

用户在浏览静态网页时，Web 服务器找到网页就直接把网页文件发送给客户端，服务器不会对网页做任何处理，静态网页的工作流程如图 1-8 所示。用户在每次浏览静态网页时，内容都不会发生变化，网页一经编写完成，其显示效果就确定了。如果要改变静态网页的内容，就必须修改网页的源代码再重新上传到服务器。

图 1-6　动态网页 1

图 1-7　动态网页 2

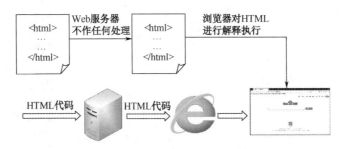

图 1-8 静态网页的工作流程

1.2.2 为什么需要动态网页

静态网页在很多时候是无法满足 Web 应用需求的。举个例子来说，假设有个电子商务网站需要展示 1000 种商品，其中每个页面显示一种商品。如果用静态网页来做的话，需要制作 1000 个静态网页，这带来的工作量是非常大的。而且以后要修改这些网页的外观风格时，就需要逐个网页进行修改，工作量也很大。如果使用动态网页来做，只需要制作一个页面，然后把 1000 种商品的信息存储在数据库中，页面会根据浏览者的请求调用数据库中的数据，即可用同一个页面显示不同商品的信息，要修改页面外观时，也只需修改这一个动态网页的外观，工作量大为减少。

由此可见，动态网页是页面中内容会根据具体情况发生变化的网页，同一个网页根据每次请求的不同，可每次显示不同的内容。例如，在一个新闻网站中，单击不同的链接可能都会链接到同一个动态页面，只是该页面能每次显示不同的新闻。

动态网页技术还能实现如留言板、论坛、博客等各种交互功能，可见动态网页带来的好处是显而易见的。动态网页要显示不同的内容，往往需要数据库做支持，这也是动态网页的一个特点。从页面的源代码看，动态网页中含有服务器端代码，需要先由 Web 服务器对这些服务器端代码进行解释执行生成 HTML 代码后，再发送给客户端。

可以从文件扩展名判断一个网页是动态网页还是静态网页。静态网页的文件扩展名是 htm、html、xml 等；动态网页的扩展名是 php、asp、jsp 等。例如 http://www.51zxw.net/list.aspx?cid=604 是一个动态网页，而 http://ec.hynu.cn/items/g1.html 是一个静态网页。

> **提示**：动态网页绝不是页面上含有动画的网页，即使在静态网页上有一些动画（如 Flash 或 gif 动画）或视频，但每次访问时，显示的内容是一样的，因此仍然属于静态网页。

习题

1. 简述 B/C 结构和 C/S 结构的区别。
2. 简述运行 Web 应用程序的要素。
3. 简述静态 Web 和动态 Web 的区别。

第 2 章

JSP 工作原理、开发环境及运行配置

JSP 作为 Java Web 开发体系中的核心技术，学习者除要了解 Web 相关的基础知识，还应该对 JSP 依赖的先行知识有所了解，如 Java 语言、Servlet 技术（Servlet 在后面章节中有详细讲解）、JSP 所开发的应用程序体系架构。下面针对这些内容进行介绍。

本章任务

（1） JSP 技术概述；
（2） JSP 在 JavaWeb 开发中的地位；
（3） JSP 开发环境搭建；
（4） JSP 开发工具；
（5） 用 IDEA 创建第一个 JSP 应用。

重点内容

（1） JSP 在 JavaWeb 开发中的地位；
（2） JSP 运行环境、安装及配置。

难点内容

创建第一个 JSP 应用。

2.1 JSP 技术概述

在了解 JSP 技术之前，我们需要先了解一些与 JSP 相关的概念，这样能帮助我们更好地学习后面的内容。

2.1.1 Java 语言

Java 语言是由 Sun 公司于 1995 年推出的一种编程语言，一经推出。就赢得了业界的一

致好评，并受到了广泛关注。Java 语言适用于 Internet 环境，目前已成为开发 Internet 应用的主要语言之一。它具有简单、面向对象、可移植性、分布性、解释器通用性、稳健、多线程、安全和高性能等优点。其中最重要的就是实现了跨平台运行，这使得应用 Java 开发的程序可以方便地移植到不同的操作系统中运行。

Java 语言是完全面向对象的编程语言，它的语法规则和 C++类似，但 Java 语言对 C++进行了简化和提高。例如，C++中的指针和多重继承通常会使程序变得复杂，而 Java 语言通过接口取代了多重继承，并取消了指针、内存的申请和释放等影响系统安全的部分。

在 Java 语言中，最小的单位是类，不允许在类外面定义变量和方法，所以就不存在所谓的"全局变量"这一概念。在 Java 类中，定义的变量和方法分别称为成员变量和成员方法，其中成员变量也叫类的属性，在定义这些类的成员时，需要通过权限修饰符来声明它们的使用范围。

Java 语言编写的程序应被保存为后缀名为.Java 的文件，然后编译成后缀名.class 的字节码文件，最终通过执行该字节码文件执行 Java 程序。

2.1.2　Servlet 技术

Servlet 是在 JSP 之前就存在的运行在服务器端的一种 Java 技术，它是用 Java 语言编写的服务器端程序，具有独立于平台和协议的特性（可在多种系统平台和服务器平台下运行）、功能强大、安全、可拓展和灵活等优点。Java 语言能够实现的功能，Servlet 基本上都可以实现（除图形页面）。Servlet 主要用于处理 HTTP 请求，并将处理的结果传递给浏览器生成动态 Web 页面。

在 JSP 中用到的 Servlet 通常都继承自 javax.Servlet.HttpServlet 类，在该类中实现了用来处理 HTTP 请求的大部分功能。

但 Servlet 有个缺陷，它不能将页面显示的代码和逻辑处理的代码进行有效的隔离，为了克服 Servlet 这一致命缺陷，人们发明了 JSP。JSP（Java Server Pages）是一种实现普通静态 XHTML 和动态 XHTML 混合编码的技术，它并没有增加任何本质上不能用 Servlet 实现的功能。所以 JSP 与 Servlet 有着密不可分的关系。JSP 页面在执行过程中会被转换为 Servlet，然后由服务器执行该 Servlet。

2.1.3　JavaBean 技术

JavaBean 是根据特殊规范编写的普通 Java 类，可称它们为"独立的组件"。每个 JavaBean 实现一个特定的功能，通过合理地组织具有不同功能的 JavaBean，可以快速地生成一个全新的应用程序。如果将这个应用程序比作一辆汽车，那么程序中的 JavaBean 就好比组成这辆汽车的不同零件。对于程序开发人员来说，JavaBean 的最大优点就是充分提高了代码的可重用性，并且对程序的后期维护和扩展起到了积极的作用。

JavaBean 可按功能划分为可视化和不可视化两种。可视化 JavaBean 主要应用在图形页面编程的领域中，在 JSP 中通常应用的是不可视化 JavaBean。应用该种 JavaBean 可用来封装各种业务逻辑，例如连接数据库、获取当前时间等。这样，当在开发程序的过程中需要连接数据库或实现其他功能时，就可直接在 JSP 页面或 Servlet 中调用实现该功能的 JavaBean 来实现。通过应用 JavaBean，可以很好地将业务逻辑和前台显示代码分离，这大大提高了代码的可读性和易维护性。

2.1.4 JSP 技术

Java Server Pages（JSP）是由 Sun 公司倡导，与多个公司共同建立的一种技术标准，它建立在 Servlet 之上。应用 JSP，程序员或非程序员可以高效率地创建 Web 应用程序，并使得开发的 Web 应用程序具有安全性高、跨平台等优点。

JSP 是运行在服务器端的脚本语言之一，与其他服务器端的脚本语言一样，是用来开发动态网页的一种技术。

JSP 页面由传统的 HTML 代码和嵌入其中的 Java 代码组成。当用户请求一个 JSP 页面时，服务器会执行这些 Java 代码，然后将结果与页面中的静态部分相结合返回给客户端浏览器。JSP 页面中还包含了各种特殊的 JSP 元素，通过这些元素可以访问其他的动态内容并将它们嵌入页面，例如访问 JavaBean 组件的<jsp:useBean>动作元素。程序员还可以通过编写自己的元素来实现特定的功能，开发出更为强大的 Web 应用程序。

JSP 是在 Servlet 的基础上开发的技术，它继承了 Java Servlet 的各项优秀功能。而 Java Servlet 是作为 Java 的一种解决方案，在制作网页的过程中，它继承了 Java 的所有特性。因此 JSP 同样继承了 Java 技术的简单、便利、面向对象、跨平台和安全可靠等优点，比起其他服务器脚本语言，JSP 更加简单、迅速和便利。在 JSP 中利用 JavaBean 和 JSP 元素，可以有效地将静态的 HTML 代码和动态数据区分开来，给程序的修改和扩展带来了很大方便。

2.1.5 JSP 和 Java Servlet 的关系

Java Servlet 是 Java 语言的一部分，提供了用于服务器编程的 API。Java Servlet 就是编写在服务器端创建对象的 Java 类，习惯上称为 Servlet 类，Servlet 类的对象习惯上称为一个 Servlet。在 JSP 技术出现之前，Web 应用开发人员就是自己编写 Servlet 类，并负责编译生成字节码文件、复制这个字节码文件到服务器的特定目录中，以便服务器使用这个 Servlet 类的字节码创建一个 Servlet 来响应用户的请求。Java Servlet 的最大缺点是不能有效地管理页面的逻辑部分和页面的输出部分，导致 Servlet 类的代码非常混乱，单独用 Java Servlet 来管理网站变成一件很困难的事情。为了克服 Java Servlet 的缺点，Sun 公司用 Java Servlet 作为基础，推出了 Java Server Page。当客户请求一个 JSP 页面时，Tomcat 服务器自动生成 Java 文件（如 first1$jsp.java）、编译 Java 文件，并用编译得到的字节码文件在服务器端创建一个 Servlet。但是 JSP 技术不是 Java Servlet 技术的全部，它只是 Java Servlet 技术的一个成功应用。JSP 技术屏蔽了 Servlet 创建的过程，使得 Web 程序设计者只需关心 JSP 页面本身的结构、设计好各种标记，比如使用 HTML 标记设计页面的视图，使用 JavaBean 标记有效地分离页面的视图和数据存储等。

有些 Web 应用可能只需要 JSP+JavaBean 就能设计得很好，但是，对于某些 Web 应用，就可能需要 JSP+JavaBean+Servlet 来完成，即需要服务器再创建一些 Servlet 对象，配合 JSP 页面来完成整个 Web 应用程序的工作。

2.1.6 JSP 在 JavaWeb 开发中的地位

Web 应用程序大体上可以分为两种，即静态网站和动态网站。早期的 Web 应用主要是静态页面的浏览，即静态网站。这些网站使用 HTML 语言来编写，放在 Web 服务器上，用户使用浏览器通过 HTTP 协议请求服务器上的 Web 页面；服务器上的 Web 服务器将接收到的用户请求处理后，再发送给客户端浏览器，显示给用户。

在开发 Web 应用程序时，通常需要应用客户端和服务器两方面的技术。其中，客户端应用的技术主要用于展现信息内容；而服务器应用的技术，则主要用于进行业务逻辑的处理和数据库的交互。想要开发动态网站，就需要用到服务器端技术。而 JSP 技术就是服务器端应用的技术，JSP 页面中的 HTML 代码用来显示静态内容部分，嵌入页面中的 Java 代码与 JSP 标记用来生成动态的内容部分。JSP 允许程序员编写自己的标签库来完成应用程序的特定要求。JSP 可以被预编译，提高了程序的运行速度。另外，JSP 开发的应用程序经过一次编译后，便可以随时随地运行。所以在绝大部分系统平台中，代码无须修改便可支持 JSP 在任何服务器中运行。

2.2 JSP 工作原理

当客户端浏览器向服务器发出请求要访问一个 JSP 页面时，服务器根据该请求加载相应的 JSP 页面，并对该页面进行编译，然后执行。JSP 的处理过程如图 2-1 所示。

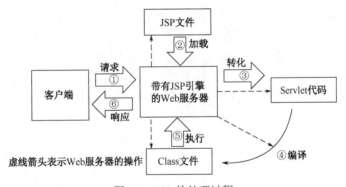

图 2-1 JSP 的处理过程

（1）客户端通过浏览器向服务器发出请求，在该请求中包含了请求资源的路径，这样当服务器接收到该请求后就可以知道被请求的资源。
（2）服务器根据接收到的客户端请求来加载被请求的 JSP 文件。
（3）Web 服务器中的 JSP 引擎会将被加载的 JSP 文件转化为 Servlet。
（4）JSP 引擎将生成的 Servlet 代码编译成 Class 文件。
（5）服务器执行这个 Class 文件。
（6）最后服务器将执行结果发送给浏览器进行显示。

从上面的介绍中可以看到，JSP 文件被 JSP 引擎进行转换后，又被编译成了 Class 文件，最终由服务器通过执行这个 Class 文件来对客户端的请求进行响应。其中（3）与（4）构成了 JSP 处理过程中的翻译阶段，而（5）为请求处理阶段。

但并不是每次请求都需要重复进行这样的处理。当服务器第一次接收到对某个页面的请求时，JSP 引擎就开始进行上述的处理过程，将被请求的 JSP 文件编译成 Class 文件。在后续对该页面再次进行请求时，若页面没有进行任何改动，服务器只需直接调用 Class 文件执行即可。所以当某个 JSP 页面第一次被请求时，会有一些延迟，而再次访问时会感觉快了很多。如果被请求的页面经过修改，服务器将会重新编译这个文件，然后执行。

2.3 JSP 开发环境搭建

2.3.1 JSP 的运行环境

使用 JSP 进行开发时，需要具备以下对应的运行环境：Web 浏览器、Web 服务器、JDK 开发工具包及数据库。下面分别介绍这些环境。

1. Web 浏览器

浏览器主要用于客户端用户访问 Web 应用工具，与开发 JSP 应用不存在很大的关系，所以开发 JSP 对浏览器的要求并不是很高，任何支持 HTML 的浏览器都可以，如 IE 浏览器、火狐浏览器、谷歌浏览器等。

2. Web 服务器

Web 服务器是运行及发布 Web 应用的大容器，只有将开发的 Web 项目放置到该容器中，才能使网络中的所有用户通过浏览器进行访问。开发 JSP 应用所采用的服务器主要是 Servlet 兼容的 Web 服务器，比较常用的有 IBM WebSphere、Apache Tomcat 和 BEA WebLogic 等。

Tomcat 服务器最为流行，它是 Apache-Jarkarta 开源项目中的一个子项目，是一个小型的、轻量级的、支持 JSP 和 Servlet 技术的 Web 服务器，已经成为学习开发 JSP 应用的首选。本书中的所有例子都使用了 Tomcat 作为 Web 服务器，所以对该服务器的安装与配置在后面的章节中也进行了讲解。

3. JDK

JDK（Java Develop Kit，Java 开发工具包）包括运行 Java 程序所必需的 JRE 环境及开发过程中常用到的库文件。在使用 JSP 开发网站之前，首先必须安装 JDK。

4. 数据库

任何项目的开发几乎都需要使用数据库，数据库用来存储项目中需要的信息。根据项目的规模采用合适的数据库。如大型项目可以采用 Oracle 数据库，中型项目可采用 Microsoft SQL Server 或 MySQL 数据库，小型项目可采用 Microsoft Access 数据库。Microsoft Access 数据库的功能远比不上 Microsoft SQL Server 和 MySQL 强大，但它具有方便、灵活的特点，对于一些小型项目来说是比较理想的选择。

2.3.2　JDK 的安装与配置

1. 下载 JDK

本书所有范例都是基于 JDK 8 开发的，所以我们下载 JDK 8，可以在 Oracle 公司的官方网站下载到。Java SE8 对应版本下载如图 2-2 所示。

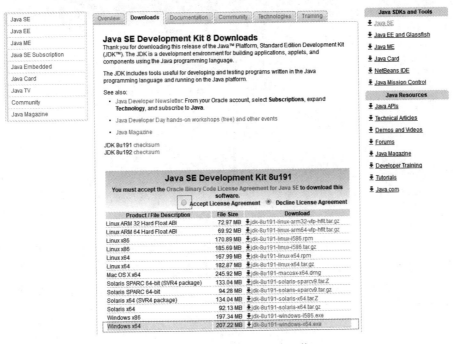

图 2-2　Java SE8 对应版本下载

2. 安装 JDK

具体安装步骤如下：

（1）双击安装文件，在弹出的欢迎页面中，单击"下一步"按钮，如图 2-3 所示。

（2）在选择安装路径的时候，我们这里使用默认的安装路径，如果你想更改默认路径，单击"更改"按钮，这里直接单击"下一步"按钮开始安装，如图 2-4 所示。

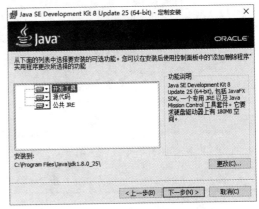

图 2-3　欢迎页面　　　　　　　　　　图 2-4　安装路径

（3）在安装的过程中，会弹出另一个"Java 安装"对话框，提示用户选择 Java 运行时环境的安装路径。这里依旧使用默认的安装路径。单击"下一步"按钮，如图 2-5 所示。

（4）等待片刻，弹出安装成功对话框。这样我们就安装好了 JDK。单击"关闭"按钮，完成安装，如图 2-6 所示。

图 2-5　Java 运行环境的路径

图 2-6　安装完成

3. 配置环境变量

在 Windows XP/ Windows 8 / Windows 10 /Windows Server 2003 的桌面上，右击"我的电脑"图标，在弹出的快捷菜单中选择"属性"命令。在"系统属性"对话框中选择"高级"选项卡，在其中单击"环境变量"按钮，如图 2-7 所示。在"环境变量"对话框（图 2-8）中新建表 2-1 所示的变量名和变量值。

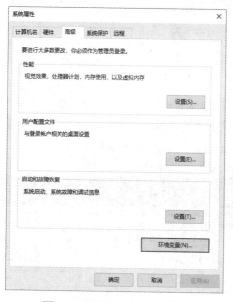

图 2-7　选择"高级"选项卡

图 2-8　设置环境变量

表 2-1　JDK 环境变量

变 量 名	变 量 值	功　　能
JAVA_HOME	C:\Program Files\Java\jdk1.8.0_25	说明 JDK 所在的搜索功能

续表

变量名	变量值	功 能
Path	C:\Program Files\Java\jdk1.8.0_25\bin 或%JAVA_HOME%\bin	说明 Java 使用程序的位置
CLASSPATH	.;%JAVA_HOME%\lib\dt.jar; %JAVA_HOME%\lib\tools.jar; %JAVA_HOME%\jre\lib\rt.jar	说明类和包文件的搜索路径

具体操作步骤如下:

① 在"lyp 的用户变量"选项区域中单击"新建"按钮,在弹出的"新建用户变量"对话框中输入变量名 JAVA_HOME 和变量值,如图 2-9 所示。

图 2-9 设置用户变量

② 在"系统变量"选项区域中双击"Path"变量,在弹出的"新建系统变量"对话框的"变量值"文本框中添加 Path 变量值,如图 2-10 所示。

③ 在"lyp 的用户变量"选项区域中单击"新建"按钮,在弹出的"新建用户变量"对话框中输入 CLASSPATH 变量名和变量值,如图 2-11 所示。

图 2-10 添加 Path

图 2-11 添加 CLASSPATH

④ 最后检验是否配置成功。在命令提示符下运行"cmd"命令,输入 java-version,可以看到安装 JDK 的版本号,如图 2-12 所示。

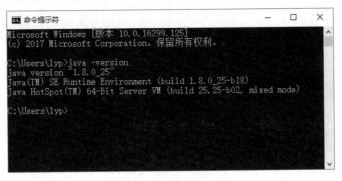

图 2-12 检验配置是否成功

2.3.3 Tomcat 7 的安装与配置

Tomcat 服务器是由 JavaSoft 和 Apache 开发团队共同提出并合作开发的产品。它能够支持 Servlet 3.0 和 JSP 2.2,并且具有免费、跨平台等诸多特性。甲骨文公司对它的支持也相当不错,Tomcat 服务器已经成为学习开发 JSP 应用的首选,本书中的所有例子都使用了

Tomcat 作为 Web 服务器。

1. 下载 Tomcat

本书中采用的是 Tomcat 7.0 版本，读者可到 Tomcat 官方网站进行下载。在下载页面中单击"32-bit/64-bit Windows Service Installer (pgp, sha1, sha512)"超链接，下载 Tomcat，如图 2-13 所示。

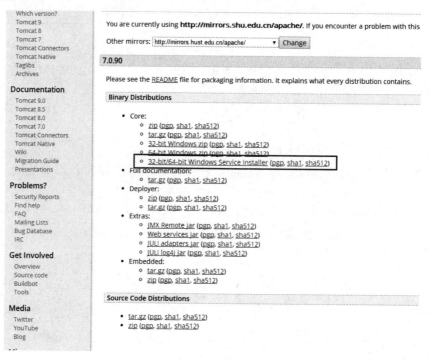

图 2-13　下载 Tomcat

2. 安装 Tomcat

下载后的文件名为 apache-tomcat-7.0.55.exe，双击该文件即可安装 Tomcat，具体安装步骤如下。

（1）双击 apache-tomcat-7.0.55.exe 文件，弹出安装向导对话框，如图 2-14 所示。单击"Next"按钮后，将弹出许可协议对话框，如图 2-15 所示。

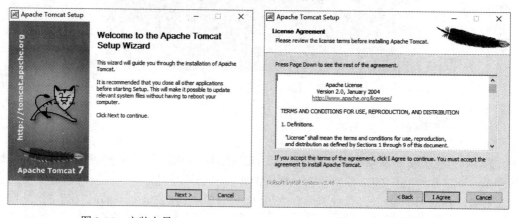

图 2-14　安装向导　　　　　　　　　　图 2-15　许可协议

（2）单击"I Agree"按钮，接受许可协议，将弹出"Choose Components"对话框，在该对话框中选需要安装的组件，通常采用其默认设置，如图 2-16 所示。

（3）单击图 2-16 中的"Next"按钮，在弹出的对话框中设置访问 Tomcat 服务器的端口、用户名和密码，采用默认设置，即端口为"8080"，用户名为"admin"，密码为空。如图 2-17 所示。

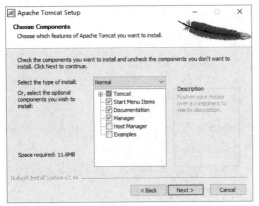

图 2-16　选择要安装的 Tomcat 组件

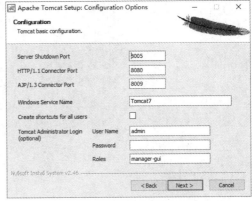

图 2-17　设置端口、用户名和密码

（4）单击"Next"按钮，在打开的 Java Virtual Machine 对话框中选择 Java 虚拟机路径，如图 2-18 所示。

（5）单击"Next"按钮，在打开的"Choose Install Location"对话框中可通过单击"Browse"按钮更改 Tomcat 的安装路径，此处我们使用默认路径，如图 2-19 所示。

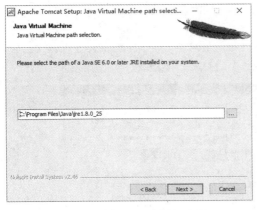

图 2-18　选择 Java 虚拟机路径

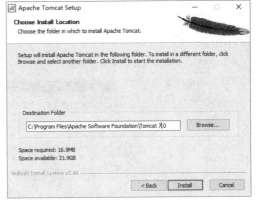

图 2-19　设置 Tomcat 安装路径

（6）单击"Install"按钮，开始安装 Tomcat。安装完成后，单击"Finish"按钮，如图 2-20 所示。

3．启动与停止 Tomcat

找到安装好的 Tomcat 服务图标，如图 2-21 所示，单击"Tomcat7w"，在打开的对话框中单击"Start"按钮后再单击"确定"按钮，如图 2-22 所示。

4．测试 Tomcat

成功安装 Tomcat 后，在浏览器中输入 http://127.0.0.1:8080 或 http://localhost:8080，如

果出现如图 2-23 所示的 Tomcat 默认主页，则表示 Tomcat 服务器配置正常。

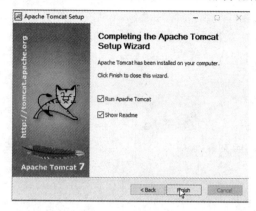

图 2-20　安装成功

图 2-21　Tomcat 图标

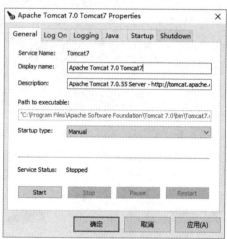

图 2-22　启动 Tomcat

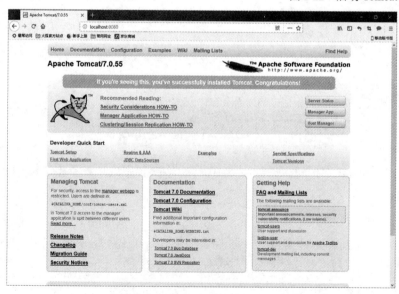

图 2-23　Tomcat 服务器配置正常

2.4 JSP 开发工具

2.4.1 IDEA 简介

IDEA 全称 IntelliJ IDEA，是用于 Java 语言开发的集成环境（也可用于其他语言），IntelliJ 在业界被公认为最好的 Java 开发工具之一，尤其在智能代码助手、代码自动提示、重构、J2EE 支持、Ant、JUnit、CVS 整合、代码审查、创新的 GUI 设计等方面的功能可以说是超常的。

2.4.2 IDEA 的安装及配置

（1）首先，在 IDEA 官网中选择我们使用的操作系统，这里使用的是 Windows 10 的 64 位系统，然后选择要下载的 IDEA 版本，有企业版和开源版两种，根据自己的需求下载相应的版本，单击"Download"按钮下载，如图 2-24 所示。

图 2-24　IDEA 下载

（2）Windows 版本的安装比较简单，找到下载好的可执行文件，然后双击，可以看到如图 2-25 所示的 IDEA 欢迎页面。

（3）单击"Next"按钮，进入选择安装目录对话框，选择要安装的位置。可以根据自己的实际情况选择安装位置，这里使用默认设置，如图 2-26 所示。

图 2-25　IDEA 欢迎页面

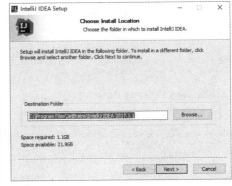

图 2-26　安装路径

（4）新版的 IDEA，提供了我们需要的操作系统位数，读者在选择操作系统的位数时，还需要安装一些插件，这里只提供了部分插件，如图 2-27 所示。

（5）接下来就是选择菜单，然后单击"Install"按钮，如图 2-28 所示。

图 2-27　选择操作系统和插件

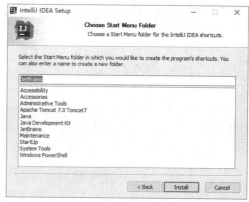
图 2-28　选择菜单

（6）安装结束后，会出现如图 2-29 所示的安装完成页面，此时会询问我们是否现在打开 IDEA，勾选 "Run IntelliJ IDEA" 复选框，单击 "Finish" 按钮。如图 2-29 所示。

（7）下面对 IDEA 进行配置，如图 2-30 所示，因为是第一次安装，所以选择下面的 "Do not import settings" 单选按钮，并单击 "OK" 按钮。

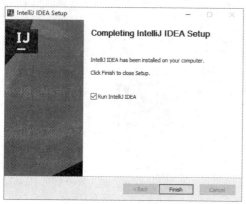

图 2-29　安装完成　　　　　　　　　　图 2-30　进行配置

（8）在激活页面单击 "Accept" 按钮，如图 2-31 所示，接着选中 "License server" 单选按钮，输入激活的网址，http://idea.iteblog.com/key.php。如果你购买了正版，则选择第一个选项 "JetBrains Account"，单击 "Activate" 按钮，激活成功，如图 2-32 所示。

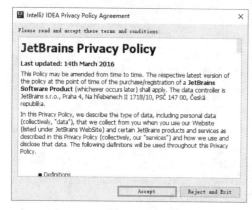

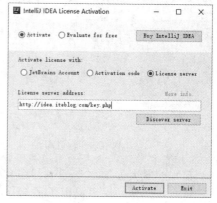

图 2-31　激活页面　　　　　　　　　　图 2-32　激活 IDEA

（9）注册完成之后，选择以后编程的页面，如图 2-33 所示。

（10）此时是配置 IDEA 支持的功能，默认有很多插件，我们可以自行选择。值得一提的是，这时配置的功能，以后还可以修改，所以不必担心，如图 2-34 所示。

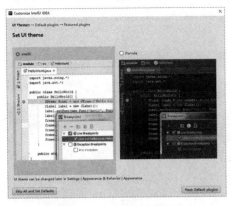

图 2-33　选择编程页面

图 2-34　选择插件

（11）安装插件，如图 2-35 所示。安装完成后，我们就可以使用 IDEA 了，其页面非常炫酷，如图 2-36 所示。

图 2-35　安装插件

图 2-36　IDEA 启动页面

2.5　创建第一个 JSP 应用

首先，必须做一个说明，在 IDEA 中有两个非常重要的概念，分别为 Project 和 Module，注意 Project 是指工作空间，而 Module 指的是工作空间下面的一个应用，如一个 Web 应用就是一个 Module（IDEA 中的 Project 相当于 MyEclipse 中的 Workspace，而 IDEA 中的 Module 相当于 MyEclipse 中的 Project）。

案例描述

创建第一个 JSP 页面，页面显示 "Hello JSP！"。

案例分析

这是个简单的例子，本例中只包含一个 index.JSP 页面，运行后，我们能在页面上看到

"Welcome To JSP Word！Hello JSP！"。从这个案例的代码中，我们能很直观地看出，JSP 页面是由 HTML 代码、JSP 元素和嵌入 HTML 代码的 Java 代码构成的。当用户请求该页面时，服务器就会加载该页面，并且会执行页面中的 JSP 元素和 Java 代码。最后将执行结果与 HTML 代码一起返回给客户端，由客户端浏览器进行显示。

案例实现

1. 新建 Web 项目

① 双击 IDEA 图标，打开 IDEA。执行"File"→"New"→"Project"命令，创建一个"New Project"，左边选择"Java"选项，右边选择 Project SDK 为 1.8，在 Java EE 下选择"Web Application（3.1）"复选框，注意窗口下的 Version 对应为 3.1，且 Create web.xml 已勾选。确定没问题后，单击"Next"按钮，如图 2-37 所示。

② 创建的新项目取名为"FirstJSP"，位置为：C:\Users\lyp\IdeaProjects\FirstJSP，这里要说明的是，C:\Users\lyp\IdeaProjects 是指 IDEA 创建的工作空间。在 IDEA 里以后我们会创建多个项目，每个项目都可以取不同的名称，但是它们都是保存在 C:\Users\lyp\IdeaProjects 这个工作空间下面的，且名称不能重复。而后面的"FirstJSP"就现在工作空间下的一个项目，也叫作 Module。这里的工作空间路径是可以根据自己的需求更改的，单击"Finish"按钮，如图 2-38 所示。

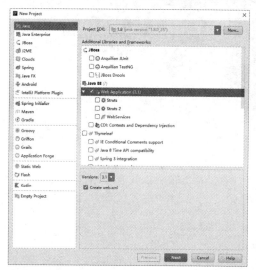 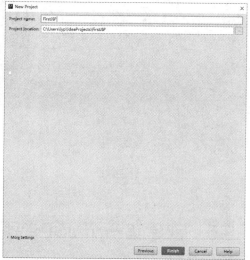

图 2-37　创建一个"New Project"　　　　图 2-38　创建第一个项目

③ 创建完成后，我们可以看到所创建的项目，如图 2-39 所示。

④ 在 web/WEB-INF 目录下创建两个文件夹：class 和 lib。选中"WEB-INF"，右击，在打开的快捷菜单中选择"New"→"Directory"命令，在弹出的对话框中输入"class"，lib 文件夹同理。class 用来存放编译后输出的 class 文件，lib 用于存放第三方 jar 包。

⑤ 执行"File"→"Project Structure"命令，在弹出的对话框中，选择"Modules"→"Paths"选项，选中"Use module compile output path"单选按钮，将"Output path"和"Test output path"都设置为刚刚创建的 class 文件夹，如图 2-40 所示。

⑥ 选择"Dependencies"，具体设置如图 2-41 所示。

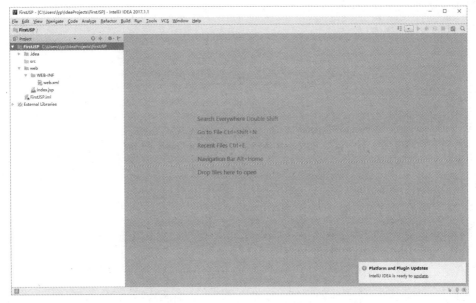

图 2-39 创建完成的第一个项目

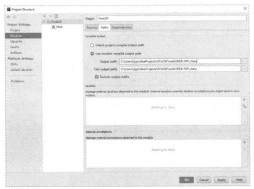

图 2-40 设置 Paths

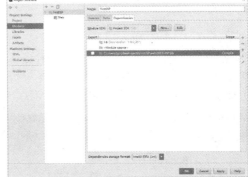

图 2-41 设置"Dependencies"

⑦ 单击"File"→"Project Structure"命令，在弹出的对话框中（图 2-42），单击右边的"+"号按钮，在下拉菜单中选择"2 Library…"，在弹出的对话框中选择"Tomcat 7.0.55"，再单击"Add Selected"按钮，如图 2-43 所示。完成 JSP 内置对象所依赖的 jar 包如图 2-44 所示。

图 2-42 选择"2 Library…"

图 2-43 选择"Tomcat 7.0.55"

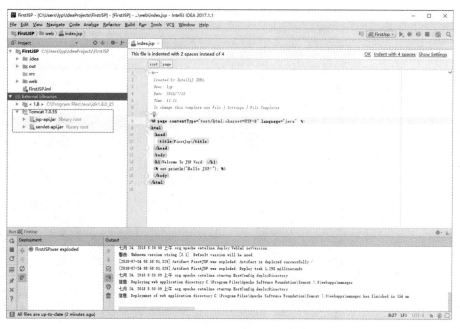

图 2-44　完成 JSP 内置对象所依赖的 jar 包

2. 配置 Tomcat 容器

① 为"FirstJSP"配置 Tomcat，执行"File"→"Settings"命令，在"Build, Execution, Deployment"下找到"Application Servers"，再单击"+"号，在弹出的对话框中，Tomcat Home 里选择安装 Tomcat 的路径，这里的路径为：C:\Program Files\Apache Software Foundation\Tomcat 7.0。选好路径后，下面的 Tomcat base directory 会自动生成路径。配置完成后单击"OK"按钮，如图 2-45 所示。

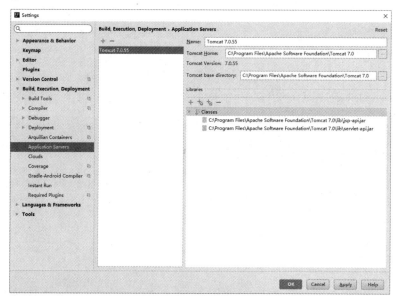

图 2-45　配置 Tomcat 容器

② 执行"Run"→"Edit Configurations…"命令，再单击"+"，选择"Tomcat Server"，

在 Name 一栏输入服务器的名字"FirstJSP",在"Server"选项卡下,取消对"After Launch"复选框的选择,配置完成后单击"OK"按钮,如图 2-46 所示。

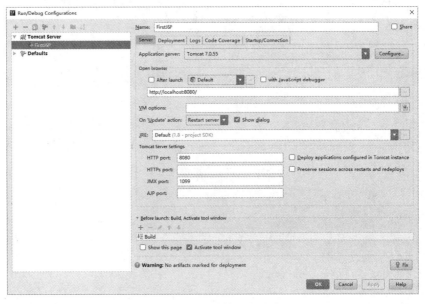

图 2-46　配置服务器的名字

3. 在 Tomcat 上部署并运行项目

① 在创建好 Tomcat 后,可以通过工具栏快速打开 Tomcat 的配置页面,也可以通过执行"Run"→"Edit Configurations..."命令,选择刚创建的"FirstJSP",单击"Deployment"选项卡,再单击右边的"+"号,在"Application context"中选择"/FirstJSP"(也可以不填,但建议与项目名称相同),打开 Tomcat 的配置页面,如图 2-47 所示。

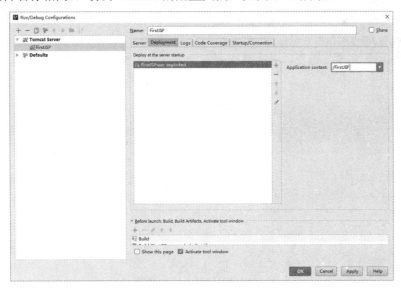

图 2-47　Tomcat 配置页面

② 返回 Server 面板,将"On 'Update' action"和"On frame deactivation"(这两个选项是 Tomcat 配置了项目后才有的)改为"Update classes and resources",如图 2-48 所示。

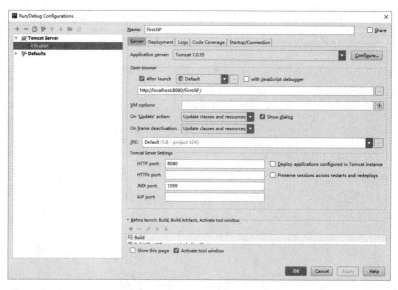

图 2-48　在 Tomcat 上部署项目

③ 双击 index.jsp 文件，将代码更改为：

```
<%@ page contentType="text/html;charset=UTF-8" language="java" %>
<html>
  <head>
    <title>FirstJSP</title>
  </head>
  <body>
  <h1>Welcome To JSP Word! </h1>
  <% out.printIn("Hello JSP!"); %>
  </body>
</html>
```

单击右上角的运行按钮，选择 FirstJSP，单击右边页面中的绿色三角形按钮运行，之后自动打开一个网址为：http://localhost:8080/FirstJSP/的 JSP 页面。如图 2-49 所示。

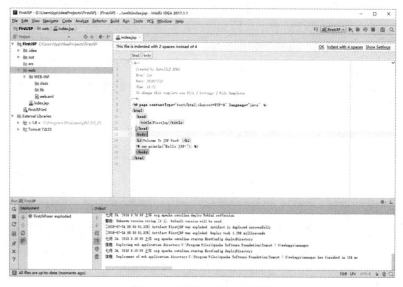

图 2-49　运行 FirstJSP 页面

运行结果

浏览器中显示的返回结果如图 2-50 所示。

图 2-50　浏览器中显示的效果

习题

1. 简述 JSP 的执行过程。
2. 学会安装与配置 IDEA。
3. 创建一个静态网页，在浏览器正中显示"Hello JSP"。

第 3 章

JSP 编程基础

在 JSP 页面结构中，除可以包含静态页面中的内容，还可以有 Java 源代码。本章详细讲解利用 Java 技术如何开发 JSP 动态页面，通过该章的学习，掌握 JSP 开发涉及的相关语法及其基本使用。

本章任务

（1）JSP 编程语法；
（2）JSP 页面编程。

重点内容

（1）JSP 页面基本的机构；
（2）脚本程序；
（3）变量和方法的声明。

难点内容

（1）脚本程序项目应用；
（2）JSP 编程语法应用。

3.1 JSP 编程语法

3.1.1 JSP 页面的基本结构

在传统的 HTML 页面文件中加入 Java 程序和 JSP 标记就构成了一个 JSP 页面。一个 JSP 页面可由以下 5 种元素组合而成。

（1）普通的 HTML 标记符。

（2）JSP 标记，如指令标记和动作标记。

（3）变量和方法的声明。

（4）Java 程序。

（5）Java 表达式。

当服务器上的一个 JSP 页面被第一次请求执行时，服务器上的 JSP 引擎首先将 JSP 页面文件转译成一个 Java 文件，再将这个 Java 文件编译生成字节码文件，然后通过执行字节码文件响应用户的请求。当多个用户请求一个 JSP 页面时，JSP 引擎为每个用户启动一个线程，该线程负责执行常驻内存的字节码文件来响应用户的请求。这些线程由 Tomcat 服务器来管理，将 CPU 的使用权在各个线程之间快速切换，以保证每个线程都有机会执行字节码文件。这个字节码文件的任务就是：

（1）把 JSP 页面中普通的 HTML 标记符号，交给用户端的浏览器执行显示。

（2）JSP 标记、变量和方法声明、Java 程序由服务器负责处理和执行，将需要显示的结果发送到用户端的浏览器。

（3）Java 表达式由服务器负责计算，并将结果转化为字符串，然后交给用户端的浏览器负责显示。

下面通过一个例子来了解 JSP 页面包含的 5 种元素，效果如图 3-1 所示，其中使用 HTML 语言的<font>标记让网页显示字体，使用 Java 程序显示服务器端的时间，使用 Java 表达式显示一些变量的值。

```jsp
<%@ page contentType="text/html;charset=UTF-8" language="java" %><!-- JSP指令标记 -->
<%@ page import="java.util.Date"%>             <!-- JSP指令标记 -->
<%!
    Date date;                                  //数据声明
    public int sum(int start,int end){          //定义方法
        int sum=0;
        for(int i=start;i<=end;i++)
            sum=sum+i;
        return sum;
    }
%>
<html>                                          <!-- html标记 -->
<head>                                          <!-- html标记 -->
    <title>JSP页面的基本结构</title>              <!-- html标记 -->
</head>
<body>                                          <!-- html标记 -->

<font size="4">
    <p>
        程序创建Data对象：
        <%
            date=new Date();
            out.printIn("<br/>"+date);
            int start=1;
            int end=100;
            int sum=sum(start,end);
        %>
```

```
            <br/>从
             <%=start%>                                  <!-- Java表达式 -->
              到
              <%=end%>
             的连续和是
             <%=sum%>
        </p>
      </font>
    </body>
    </html>
```

上述代码的运行结果如图 3-1 所示。

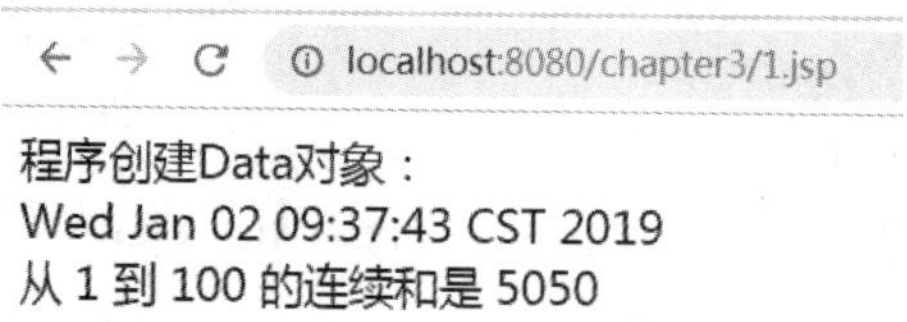
localhost:8080/chapter3/1.jsp
程序创建Data对象：
Wed Jan 02 09:37:43 CST 2019
从 1 到 100 的连续和是 5050

图 3-1　JSP 页面基本结构运行结果

3.1.2　JSP 变量的声明

在“<%!”和“%>”标记符之间声明变量，即在“<%!”和“%>”之间放置 Java 的变量声明语句。变量的类型可以是 Java 语言允许的任何数据类型，将这些变量称为 JSP 页面的成员变量。例如：

```
<%!
   int a,b=10,c;
   String tom=null,jerry="love jsp";
   Date date;
%>
```

在“<%!”和“%>”之间声明的变量在整个 JSP 页面内都有效，与“<%!”和“%>”标记符在 JSP 页面中所在的书写位置无关，但是习惯上把“<%!”和“%>”标记符写在 JSP 页面的前面。JSP 引擎将 JSP 页面转译成 Java 文件时，将“<%!”和“%>”标记符之间声明的变量作为类的成员变量，这些变量占有的内存空间直到 JSP 引擎关闭才释放。当多个用户请求 JSP 页面时，JSP 引擎为每个用户启用一个线程，这些线程由 JSP 引擎来管理，这些线程共享 JSP 页面的成员变量，因此任何一个用户对 JSP 页面成员变量操作的结果，都会影响其他用户。

下面通过例子来利用成员变量被用户共享这一性质，实现一个简单的计数器，效果如图 3-2 所示。

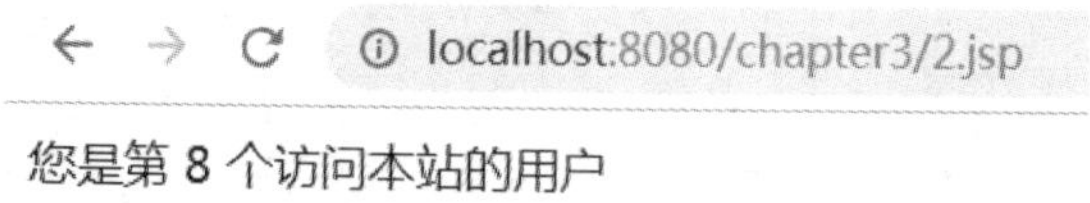
localhost:8080/chapter3/2.jsp
您是第 8 个访问本站的用户

图 3-2　声明变量

```
<%@ page contentType="text/html;charset=UTF-8" language="java" %>
<html>
<head>
```

```
        <title>变量声明</title>
    </head>
    <body>
        <%!
            int i=0;
        %>
        <%
            i++;
        %>
        <p>
            您是第
            <%=i%>
            个访问本站的用户
        </p>
    </body>
</html>
```

上述代码的运行结果如图 3-2 所示。

3.1.3 选择语句

选择语句，也称分支语句，它有特定的语法规则，需要对一些条件做出判断，条件判断的结果有两个，所以产生选择，按照不同的选择执行不同的代码。

If...else 块开始像一个普通的小脚本，但是小脚本在每行都会被闭合，并且包含小脚本标签在 HTML 文本中。实例如下：

```
<%@ page contentType="text/html;charset=UTF-8" language="java" %>
<html>
<head>
<meta http-equiv="Content-Type" content="text/html; charset=UTF-8">
<title>IF...Else示例</title>
</head>
<body>
    <%!int day = 3;%>
    <%
        if (day == 1 | day == 7) {
    %>
<p>今天是周末，不用上班，Oyes ~</p>
    <%
        } else {
    %>
<p>今天是工作日，乖乖去上班 ~</p>
    <%
        }
    %>
</body>
</html>
```

3.1.4 循环语句

在 Java 中使用三种基本类型的循环块：for，while 及 do...while 均可在 JSP 编程中使用。下面以 for 循环语句写出以下例子。

```
<%@ page language="java" contentType="text/html; charset=UTF-8"
    pageEncoding="UTF-8"%>
<!DOCTYPE html PUBLIC "-//W3C//DTD HTML 4.01 Transitional//EN" "http://
www.w3.org/TR/html4/loose.dtd">
<html>
<head>
<meta http-equiv="Content-Type" content="text/html; charset=UTF-8">
<title>For循环语句示例</title>
</head>
<body>
    <%!int fontSize;%>
    <%
        for (fontSize = 1; fontSize <= 5; fontSize++) {
    %>
    <font color="green" size="<%=fontSize%>"> JSP教程 </font>
    <br />
    <%
        }
    %>
</body>
</html>
```

3.2 JSP 页面编程

3.2.1 脚本程序

本节介绍脚本元素的基本知识。在 "<%" 和 "%>" 标记之间放置的 Java 代码称为 Java 程序片。一个 JSP 页面可以有多个 Java 程序片。程序片中声明的变量称为程序片变量，是局部变量。程序片变量的有效范围与其声明位置有关，即从声明位置向后有效，可以在声明位置后的程序片、表达式中使用；基于此，大的程序片可以分为几个小的程序片，程序片中可以插入 HTML 标记，以便使程序片代码更具可读性。

JSP 引擎将页面翻译成 Java 文件时，将程序片中的变量作为页面 Servlet 类的 Service 方法中的局部变量处理，程序片中的 Java 语句作为 Service 方法中的语句处理，最终将页面所有程序片中的变量和语句依次转译到 Service 方法中，被 JSP 引擎顺序执行。程序片可以完成以下任务。

- 程序片可以操作页面的成员变量，页面成员变量在客户线程之间共享。
- 程序片可以调用方法，调用的方法必须是页面成员方法或内置对象的方法。
- 声明和操作程序片变量。

如果多个客户访问一个 JSP 页面，JSP 页面的程序片就会被执行多次，分别运行在不同的线程中。程序片变量不同于在 "<" 和 "%" 之间声明的页面成员变量，不能在不同客户访问页面的线程之间共享，也就是运行在不同客户线程的 Java 程序片变量互不干扰；当一个客户线程执行完程序片后，该线程的程序片变量内存空间随即释放。

3.2.2 变量与方法的声明

变量和方法在"<%!"和"%>"标记之间声明,声明变量与方法的语法格式同 Java 语言,格式如下:

```
<%! declarations %>
```

1. 变量声明

声明变量就是在"<%!"和"%>"标记之间放置 Java 的变量声明语句。变量的数据类型可以是 Java 的任何数据类型。例如:

```
<%! Date datetime
    int countNum;
%>
```

dateTime 和 countNum 就是声明的变量。在"<%!"和"%>"之间声明的变量又称为页面成员变量,其作用范围为整个 JSP 页面,与书写位置无关,但一般放在页面的前面,JSP 引擎转译页面时,将"<%!"和"%>"标记之间声明的变量作为类成员变量来处理,变量占用的内存空间直到 JSP 引擎关闭时才释放;当多个客户访问同一个页面时,JSP 引擎为客户创建的线程之间共享页面成员变量,每个客户线程对页面成员变量的操作都会影响它的值。

2. 方法声明

方法声明就是在"<%!"和"%>"标记之间放置 Java 的方法声明"<%!"和"%>"之间声明的方法,在整个页面内有效,称为页面的成员方法;页面成员方法在 Java 程序片中被调用,在方法内声明的变量称为局部变量,只在方法内有效,调用完毕释放变量空间。

3.2.3 JSP 表达式

在"<%="和"%>"标记之间放置 Java 表达式,可以直接输出 Java 表达式的值。表达式的值由服务器负责计算,并将计算值转换成字符串发送给客户端显示。表达式在 JSP 编程中较常用,特别是在与 HTML 标记混合编写时使用较多。

3.2.4 JSP 中的注释

适当的注释可以增强程序的可读性,它可以方便程序的调试和维护。程序员在程序中书写注释是程序员一个良好的习惯,读者在学习时应该注意这些习惯的养成,为成为一个优秀的程序员做好准备。

(1) 输出型注释。

输出型注释是指会被 JSP 引擎发送给客户端浏览器的注释,这种注释可以在浏览器的源码中看到,浏览器将其作为 HTML 的注释处理。输出型注释的内容写在"<!--"和"-->"之间,格式如下:

```
<!-- 注释内容[<%=表达式%>]-->
```

例如:

```
<!-- 下面是Java的程序片 -->
```

在客户端的 HTML 源码中为"<!-- 下面是 Java 的程序片 -->"。

输出型注释还可以在注释内容中加入表达式,例如:

```
<!-- 页面加载时间:<%=( new java.util.Date()).toLocaleString()%>-->
```

在客户端的 HTML 源码中为"`<!-- 页面加载时间:20:08:00 -->`"。

（2）隐藏型注释。

在标记"`<%--`"和"`--%>`"之间加入的内容称为隐藏型注释，它们会被 JSP 引擎忽略，不会发送到客户端浏览器中，所以称为隐藏型注释。其使用格式如下：

```
<%--注释内容--%>
```

隐藏型注释一般写在 Java 程序片的前面，对程序片做出说明。

读者需要注意，在 Java 程序片中可以使用 Java 语言的注释方法，例如：

```
//注释内容
*注释内容*
*注释内容*
```

这些注释内容都会被 JSP 引擎忽略，不会发送到客户端浏览器中。

3.3 输出 26 个英文字母

通过输出 26 个英文字母让学生掌握程序片和表达式的应用。

案例描述

本案例具体要求如下：
- 分两行输出 26 个英文字母，每行的英文字母颜色不同。
- 要求运用到表达式输出内容。

案例分析

根据案例描述，采用 JSP 技术实现该案例。

案例实现

根据案例分析，设计的页面代码如下：

```jsp
<%@ page contentType="text/html;charset=UTF-8" language="java" %>
<%--jsp指令--%>
<html>
<head>
    <title>用不同颜色输出26个英文字母</title>
</head>
<%--HTML标记--%>
<body bgcolor="cyan">
<font size="4">
    <%
    /*Java程序片*/
        char begin = 'A';
        int ix = 13;
    %>
    <font color="blue">
        <p>蓝色输出前13个字母： </p>
```

```
<%--插入HTML标记--%>
        <%
/*Java程序片*/
        for (; begin < 'A' + ix; begin ++){
        %>
        <%= begin %>
<%--Java表达式--%>
        <%
            }
        %>
    </font>
    <font color="green">
        <p>绿色输出后13个字母：</p>
<%--插入HTML标记--%>
        <%
        for (; begin < 'N'+ ix; begin ++){
        %>
        <%= begin %>
        <%
            }
        %>
    </font>
</font>

</body>
</html>
```

运行结果

启动运行案例中的项目，运行结果如图 3-3 所示。

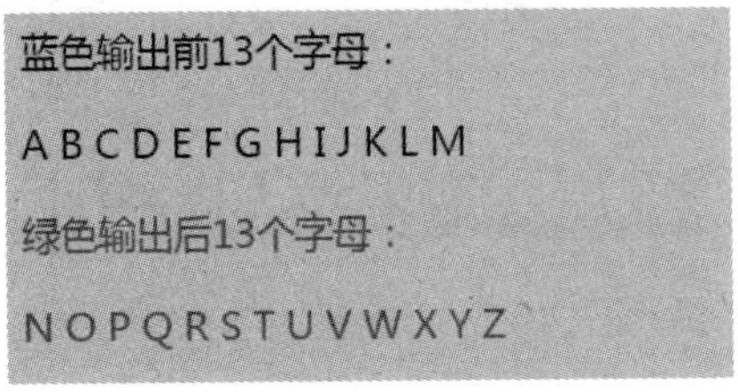

蓝色输出前13个字母：

ABCDEFGHIJKLM

绿色输出后13个字母：

NOPQRSTUVWXYZ

图 3-3　输出 26 个英文字母

3.4　抽奖游戏

本案例的抽奖游戏规则是幸运数等于总访问次数与 10 的余数。

案例描述

每个访问者随机抽取幸运数字，如果幸运数等于总访问次数与 10 的余数，则为幸运访问者。案例要求如下：
- 写出幸运数的算法。
- 要求使用 JSP 程序片。
- 显示第几次访问或者幸运者。

案例分析

根据案例描述，采用 JSP 技术实现该案例。

案例实现

根据案例分析，设计的页面代码如下：

```
<%@ page contentType="text/html;charset=UTF-8" language="java" %>
<%--jsp指令--%>
<%!
    int lucknum = 0, count = 0;
    public synchronized void countNum(){
        count ++;
        lucknum = count % 10;
    }
%>
<html>
<head>
    <title>抽奖程序示例</title>
</head>
<body bgcolor="cyan">
    <font size="4">
        <%
            countNum();
            int num = (int) (Math.random() * 10) + 1;
            if (num == lucknum){
        %>
            <p>您访问的幸运数是：<%= num %>
        <%
            }else{
        %>
            <p>您抽取的数字是：<%= num %>
            <p>您是第<%= count %>个访问者
        <%
            }
        %>
    </font>
</body>
</html>
```

运行结果

启动运行案例中的项目，运行结果如图 3-4 所示。

您抽取的数字是：7

您是第9个访问者

您访问的幸运数是：6

图 3-4 抽奖游戏

习题

1．编写一个对 1 和 100 之间的整数求和的 JSP 程序。
要求：
（1）在程序中对语句进行说明。
（2）说明采用两种注释方式。
2．请编写一个简单的 JSP 页面，显示大写英文字母表。
3．"<%!"和"%>"之间声明的变量与"<%"和"%>"之间声明的变量有何不同？
4．如果有两个访问用户访问一个 JSP 页面，该页面中的 Java 程序片将被执行几次？

第 4 章

JSP 指令操作

JSP 指令（directive）是为 JSP 引擎而设计的，它们并不直接产生任何可见输出，而只是告诉引擎如何处理 JSP 页面中的其余部分。指令用来声明 JSP 页面的一些属性，如编码方式，文档类型。在 Servlet 中也会声明我们使用的编码方式和响应的文档类型，而 JSP 就是用指令来声明的。

本章任务

（1）编译指令的掌握；
（2）动作指令的掌握；
（3）指令项目应用。

重点内容

（1）掌握编译指令；
（2）掌握动作指令；
（3）指令项目的应用。

难点内容

（1）指令的使用；
（2）指令项目应用。

4.1 编译指令

JSP 指令元素主要用来与 JSP 引擎进行沟通，它们并不会直接产生任何可见的输出；相反，它们是在告诉引擎如何处理其他的 JSP 网页。指令元素的语法格式如下：

```
<%@ 指令名 属性="属性值"%>
```

并且还可以在一个指令中加入多个属性,如:

```
<%@ 指令名 属性1="属性值1" 属性2="属性值2"
……
属性n="属性值n" %>
```

JSP 指令元素主要有三种类型的指令:

- page
- include
- Taglib

4.1.1　page 指令

page 指令描述了和页面相关的指示信息。在一个 JSP 页面中,page 指令可以出现多次,但是每种属性却只能出现一次,重复的属性设置将覆盖先前的设置。page 指令具有下面的通用格式:

```
<% @ page  language="脚本语言"
        extends="类名"
        import="Java包列表"
        session="true | false"
        buffer="none | 8KB | 自定义缓冲区大小"
        inThreadSafe="true | false"
        info="页面信息"
        errorPage="页面出错时,错误处理页面的URL"
        isErrorPage="true | false"
        contentType="内容类型信息" %>
```

接下来将对其中每个属性进行介绍。

1. language 属性

language 属性表示页面所使用的脚本语言。JSP1.1 只支持默认的 Java 语言,因此不必指定这个属性。如果设置 language 属性,则可以像下面这样来设置:

```
<% @ page language="java" %>
```

当 language 属性被设置为 java 时,该页面中所有的脚本代码必须符合 Java 程序设计语言的语法规范。

2. extends 属性

extends 属性定义了由 JSP 页面产生的 Servlet 的父类,它包含类名和所在包名的完整类名。一般来说,这个属性不会用到。但高级 JSP 开发人员可以使用该属性为其 JSP 页面创建自定义的父类。这样的父类必须实现 javax ServletHttp JspPage 接口。下面的 page 指令使当前 JSP 页面继承 myPackage 包中的 Example:

```
<%@ page extends= "myPackage . Example" %>
```

3. import 属性

import 属性用来描述哪些类可以在脚本元素中使用,属性值可以是类别全名:

```
<%@ page import="java,util.Date" %>
```

属性值也可以是包名后续接 ".*" 字符串,表示可以使用定义在这个包中的所有公共类,例如:

```
<%@ page import="java.util.*" %>
```

在 Java 语言中，如果要同时使用多个包中的类时，需要多个 import 导入多个包，例如：

```
import java.util.*;
import.java.awt.*;
```

为了实现相同的功能，在 JSP 页面中可以使用逗号为分隔符，为 import 属性赋多个值，如下所示。

```
<%@ page import="java.util.* , java.awt..*" %>
```

4. session 属性

session 属性默认值为 true，也可以为 false。该属性指定一个页面是否加入会话期的管理。如果为 true，表明内建对象 session 存在，或者可以重新产生一个 session 对象；如果为 false，表示内建对象 session 不存在，任何使用到 session 的语句都会产生 JSP 编译错误。

5. buffer 属性

buffer 属性的默认值为"8KB"，还可以是"none"或一个指定的数值。下面是一个设置缓冲区大小为 12KB 的示例。

```
<%@ page buffer-"12KB" >
```

6. autoFlush 属性

autoFlush 属性的默认值为 true，也可以改变其值为 false。它用来表明在缓存区已满时是否要自动清空。如果值为 true，则自动清空缓冲区。如果设置此属性值为 false，则当缓存区满时将产生一个 IOException 异常。

需要注意的是，当 buffer 属性值为 none 时，不能将 autoFlush 属性设为 false。通常在同一个 page 指令中设置 buffer 和 autoFalush 属性。如下所示。

```
<%@ page buffer="12KB" autoFLush="true" %>
```

7. inThreadSafe 属性

inThreadSafe 属性默认值为 true，如果多个客户请求发向 JSP 引擎时，可以被一次处理。为了同时处理多个请求，必须在 JSP 网页上编写同步处理多个请求的程序代码。如果 inThreadSafe 被设置成 false，则采用单线程模式控制客户端访问该页面。

8. info 属性

info 属性定义一个可以通过 Serviet.getServerletInfo()方法获取的信息。通常情况下，该方法返回作者、版本或版权这样的信息。

9. errorPage 属性

errorPage 属性的取值给出一个 JSP 文件的相对路径。当 errorPare 属性被定义后，如果 JSP 网页出现异常，则该异常将由 errorPage 属性指定的 JSP 网页处理。

10. isErrorPage 属性

isErrorPage 属性用来指定目前的 JSP 网页是否是另一个 JSP 网页的异常处理页，通常与 errorPage 属性配合使用。如果设置其值为 true，那么在这个 JSP 页面中可以使用内建对象 exception，以处理另一个 JSP 网页所产生的异常。如果值为 false，则不能使用内建对象 exception，否则将产生 JSP 编译错误。其默认值为 false。

11. contentType 属性

contentType 属性用来指定 JSP 网页输出到客户端时所用的 MIME 类型和字符集，可以使用任何合法的 MIME 类型和字符集。默认 MIME 类型为 text/html，默认的字符集是 ISO -8859-1 这一项须在文件的前面部分，在文件任何其他字符出现之前。如要输出简体中文，字符需要被设置为"gb2312"，如下所示。

```jsp
<%@ page contentType="text/html;charset=gb2312" %>
```

为了理解 page 指令在页面中的应用，这里演示 JSP 的一个简单的错误处理机制，首先，创建名为 rofeti 页面，并保存到 Tomcat 的 mysql 目录下。

```jsp
<%@ page contentType="text/html; charset=gb2312" language="java" errorPage="errorPage.jsp" %>
<html>
<head>
<title>errorPage示例</title>
</head>
<body>
<%
String str=null;
%>
字符串str的长度为<%=str.length()%>
</body>
</html>
```

由上面的代码可知，该程序将抛出异常 NullPointerException。注意 page 指令的 errorPage 属性被设置为 errorPage.jsp 页面。

然后，创建名为 errorPage.jsp 的文件，并将该文件保存到与 errorTest.jsp 相同的目录下。errorPage.jsp 文件中的代码如下。

```jsp
<%@page isErrorPage="true" contentType="text/html;charset=gb2312" %>
<html>
<head>
<title>错误处理页面</title>
</head>
<body>
<h1>错误信息<h1>
<hr>
<center>
<h3><%=exception%></h3>
</center>
</body>
</html>
```

最后，启动浏览器加载 JSP 页面的 errorTest.jsp 页面。加载页面时应输入的 URL 如下：

```
http://localhost:8080/myjsp/errorTest.jsp
```

这时，由于 errorText.jsp 页面将抛出异常 NullPointerException，并且该页面并没有对该异常进行处理，所以该异常将由 errorPage.jsp 页面处理。

4.1.2 include 指令

在 JSP 网页中插入其他文件有两种方式：一种是使用 include 指令，另一种是随后将要介绍的<jsp:include>动作。include 指令形式如下：

```
<%@ include file="URL" %>
```

其中，URL 是相对于 JSP 网页的，一般来说，也可以是网络服务器的根目录，被包含进来的文件内容将被解析成 JSP 文本，包含的文件必须符合 JSP 语法，应该是静态文本、脚本元素、指令元素或动作元素。

下面是一个使用 include 指令的例子。

创建包含文件名为 IncludeTest.jsp，其代码如下：

```
<%@ page contentType="text/html; charset=gb2312" language="java"%>
<html>
<head>
<title>Include指令示例</title>
</head>
<body>
<%@ include file="Test.html"%>
</body>
</html>
```

该文件中仅有一条 include 指令，include 指令包含了 Test.html 文件。Test.html 文件的代码如下：

```
<% for(int i=1;i<=5;i++) { %>
<font size="<%=i%>">Hello JSP</font><br>
<%}%>
```

这段代码实现了从小到大输出一个字符串。将这两个文件都保存到 Tomcat 目录中的 myjsp 目录下。在浏览器地址栏中输入 "http://localhost:8080/myjsp/IncludeTest.jsp"。显示效果如图 4-1 所示。

图 4-1　include 实例页面

注意，即使插入的文件扩展名不是 ".jsp"，其中的 JSP 语句也同样起作用。include 指令是编译阶段的指令，即 include 所包含文件的内容是在编译的时候插入 JSP 文件的。JSP 引擎在判断 JSP 页面是否被修改时，会对 JSP 页面文件和字节码文件的最新更改日期进行比较。由于被包含的文件是在编译时才插入的，因此如果只修改了 include 文件内容，而没有对 JSP 进行修改，得到的结果将不会改变。因为 JSP 引擎会判断 JSP 页面没被改动，所以直接执行已经存在的字节码文件，而没有重新编译。

因此对于不经常变化的内容，用 include 指令是合适的，如果包含的内容是经常变化的，则需要动作元素<jsp:include>。有关动作元素将在后面详细讨论。

4.1.3　taglib 指令

taglib 指令用于指示这个 JSP 页面所使用的标签库，标签库的具体用法属于 JSP 比较高级的内容，这里先讲述一下基本语法，关于更加详细的说明将在后面章节介绍。

```
<%@ taglib uri="tagLibraryURI" prefix="tagPrefix" %
```

属性 uri 是描述标签库位置的 uri，可以是相对路径或绝对路径。prefix 属性指定了自定义标记的前缀。

4.2 动作指令

JSP 动作元素用来控制 JSP 容器的动作，可以动态插入文件、重用 JavaBean 组件、导向另一个页面等。可用的标准动作元素有：

- <jsp:useBean>
- <jsp:setProperty>
- <jsp:getProperty>
- <jsp:include>
- <jsp:forward>
- <jsp:plugin>
- <jsp:param>

动作元素与指令元素不同，动作元素是在客户端请求时动态执行的，每次有客户端请求时可能都会被执行一次；而指令元素是在编译时被编译执行的，它只会被编译执行，且只会被编译一次。

4.2.1 include 指令

<jsp:include>动作允许将静态 HTML 或其他 JSP 的内容输出到当前 JSP 页面中。<jsp:include>动作具有两种形式。最简单的形式是不设置任何参数。其语法格式如下：

```
<jsp:include page="URL"/>
```

另一种复杂的形式支持为<jsp:param>动作设置参数。其语法格式如下：

```
<jsp:include page="URL">
 [<jsp:param./>]
</sp:include>
```

4.2.2 useBean 指令

JSP 可以动态使用 JavaBean 组件来扩充 JSP 的功能，由于 JavaBean 在开发上以及<jsp:useBean>在使用上的简单明了，使得 JSP 的开发过程和以往其他动态网页开发有了本质上的区别。

<jsp:useBean>动作的语法格式如下：

```
<jsp:useBean id="name" scope="page | request | session |application"
 typespec/>
typeSpec::= class="className" |
class="className" type="typeName" |
beanName="beanName" type="typeName" |
```

```
type="typeName"
```

各属性的含义如下：

id 属性给一个变量命名，此变量将指向 bean。在 JSP 页面中将使用这个名称访问 JavaBean。

scope 属性指定 JavaBean 组件的作用域。默认值为 page 作用域，表明了此 bean 只能应用于当前页；request 属性值表明此 bean 只能应用于当前的用户请求中；session 属性值表明此 bean 显示 JavaBean 能应用到整个 session 的生命周期；application 属性值表明此 bean 能应用于当前整个 Web 应用的范围内。

class 属性定义 JavaBean 实现的 Java 类的完全限定名称。如果没有指定 type 属性，那么必须指定 class 属性，如果指定了 beanName 属性，则不能指定 class 属性。

beanName 属性赋予 bean 一个名字，应该在 beans 的实例化方法中提供。它允许只给出 type 和 beanName 属性，而省略 class 属性。

type 属性指明将指向对象的变量的类型，变量的名由 id 属性来指定。

4.2.3　setPoperty 指令

<jsp:setProperty>动作用于为一个 JavaBean 的属性赋值。例如，可以像上一节那样定义一个 Person 类的实例。

```
<jsp:useBean id="personInfo" scope="page" class="Test.Person" />
```

然后，在这个 Person 实例上设置 name 属性。

<jsp:setPropety>的完整语法如下：

```
<jsp:setProperty name="beanName" prop_expr />
prop_expr::=property="*" |
property="propertyName" |
property="propertyName" param="parameterName" |
property="propertyName" value="propertyValue"
propertyvalue ::=string | JSP expression
```

<jsp:setProperty>的 4 个属性及其解释如下：

name：代表了需要设置属性的 JavaBean 实例名称。

property：表明了需要设定值的 JavaBean 属性名。这里有一个特殊的 property 设定，当一个 property 设定为 "*" 时，JSP 引擎将把系统 ServletRequest 对象中的参数逐个列举出来，检查这个 JavaBean 的属性是否和 ServletRequest 对象中的参数有相同的名称。如果有，就自动将 ServletRequest 对象中同名参数的值传递给相应的 JavaBean 属性。

value 这个可选属性规定了 JavaBean 实例的属性的具体值。注意，不能同时应用 value 和 param 两个属性，但允许同时不用这两个属性。该属性用来指定 Bean 属性的值。字符串数据会在目标类中通过标准的 valueOf 方法自动转换成数字、boolean、Boolean、byte、Byte、char、Character。例如，boolean 和 Boolean 类型的属性值（如 "true"）通过 Boolean.valueOf 转换，int 和 Integer 类型的属性值（如 "42"）通过 Integer.valueOf 转换。value 和 param 不能同时使用，但可以使用其中任意一个。

param 属性值是向指定的 JavaBean 属性赋值的 HTTP 请求参数名。param 是可选的。它指定用哪个请求参数作为 Bean 属性的值。如果当前请求没有参数，则什么事情也不做，系统不会把 null 传递给 Bean 属性的 set 方法。因此，你可以让 Bean 自己提供默认属性值，只有当请求参数明确指定了新值时才修改默认属性值。

4.2.4　getPoperty 指令

<jsp:getProperty>动作元素对应于 JavaBean 上设置属性的<jsp:setProperty>动作元素，<jsp:getProperty>动作用于从一个 JavaBean 中得到某个属性的值，无论原先这个属性是什么类型的，都将被转换为一个 String 类型的值。<jsp:getProperty>动作元素的语法如下：

```
<jsp:getProperty name="name" property="propertyName"/>
```

使用<jsp:getProperty>动作可以代替在 JSP 表达式内部调用方法获取属性值。例如，下面两个 JSP 代码段具有相同的结果：

```
<%=personInfo.getName() %>
<%=personInfo.getAge() %>
```

与此等价的代码如下：

```
<jsp:getProperty name="personInfo" property="name"/>
<jsp:getProperty name="personInfo" property="age"/>
```

4.2.5　forward 指令

<jsp:forward>动作用来把当前的 JSP 页面重导到另一个页面上，用户看到的地址是当前网页的地址，内容则是另一个页面的。该动作有两种形式。如果没有使用<jsp:param>动作元素添加参数，则其语法格式如下：

```
<jsp:forward page="URL"/>
```

page 属性包含的是一个相对 URL。page 的值既可以直接给出，也可以在请求的时候动态计算，可以是一个 JSP 页面或一个 Java Servlet。如果添加参数，则其语法格式如下：

```
<jsp:forward page="URL">
    [<jsp:param./>]*
</jsp:forward>"L"/>
```

<jsp.param>动作元素是配合<jsp:forward><jsp:include>和<jsp:plugin>一起使用来传递参数的。<jsp:param>动作元素的语法如下：

```
<jsp:param name="name" value="value" />
```

其中，name 表示参数名，value 表示传递的参数值。

4.2.6　plugin 指令

<jsp:plugin>动作为 Web 开发人员提供了一种在 JSP 文件中嵌入客户端运行的 Java 程序（如 Applet、JavaBean）的方法。在 JSP 处理这个动作的时候，根据客户端浏览器的不同，JSP 在执行以后将分别输出为 OBJECT 或 EMBED 这两个不同的 HTML 元素。

下面是<jsp:plugin>的语法：

```
<jsp:plugintype="bean|applet"
    type="applet | bean"
    code="classFile"
    codebase="objectCodebase"
    [align="alignment"]
    [archive="archivelist"]
    [height="height"]
```

```
        [hspace="hspace"]
        [reversion="reversion"]
        [name="componentName"]
        [vspace="vspace"]
        [width="width"]
        [nspluginurl="url"]
        [iepluginurl="url"]    >
        [ <sp:params>
          [ <sp:param name="paramName" value="paramValue" /> ]+
        </jsp:params>]
        [ <jsp:fallback> arbitrary_text </jsp:fallback> ]
    </jsp:plugin>
```

各属性及属性名称说明如表 4-1 所示。

表 4-1 各属性及属性名称说明

序 号	属 性	说 明
1	type	用来指定插件类型，可以是 Bean 和 Applet
2	name	用来指定 Applet 或 Bean 名称
3	code	用来指定所执行的 Java 类名，必须以 class 结尾
4	codebase	用来指定所执行的 Java 类所在的目录
5	align	用来指定 Applet 或 Bean 显示时的对齐方式
6	height	用来指定 Applet 或 Bean 显示时的高度
7	width	用来指定 Applet 或 Bean 显示时的宽度
8	hspace	用来指定 Applet 或 Bean 显示时距离屏幕左右的距离，单位是像素
9	vspace	用来指定 Applet 或 Bean 显示时距离屏幕上下的距离，单位是像素
10	archive	用来指定 Applet 或 Bean 执行前预先加载的类的列表
11	iepluginurl	用来指定 IE 用户能够使用的 JRE 下载地址
12	nspluginurl	用来指定 Netscape Navigator 用户能够使用的 JRE 下载地址

传递给 Applet 或 JavaBean 的参数是通过<jsp:param>动作来完成的，而<jsp:fallback>的 msg 参数是在用户浏览器不支持 Java 的情况下显示文本。

4.3 设计一个登录页面

通过登录案例让学生掌握怎么在 JSP 页面中使用指令。

案例描述

登录页面几乎是每个网站都有的，本案例具体要求如下：
- 设计登录类，类的属性分别为用户名称和密码，类的方法有设置和获取用户名称的方法、设置和获取密码的方法、判断登录是否成功的方法。
- 要求运用到<jsp:useBean><jsp:setProperty><jsp:getProperyt><jsp:forward>指令。
- 当登录成功后跳转到显示页面，登录失败后自动跳转到登录页面。

案例分析

根据案例描述，登录 LoginBeans 是一个 JavaBean 类，负责表示登录信息内容，具体设计如表 4-2 所示。

表 4-2 LoginBeans 类

类 名 称	LoginBeans	中文名称	登录	类 型	实体类
类 描 述	提供登录基本信息，属于一个 JavaBean 类				
属 性					
序 号	属 性 名	权限与修饰词	类 型	说 明	
1	loginName	private	String	登录名	
2	password	private	String	密码	
方 法					
序 号	方 法 名	参 数	返回类型	功能说明	
1	getLoginName	无	String	获取登录名名称	
2	setLoginName	String name：登录名称	void	设置登录名名称	
3	getPassword	无	String	获取密码	
4	setPassword	String color：密码	void	设置密码	
5	check	无	boolean	判断登录是否成功	

案例实现

根据案例分析，该案例由 LonginBeans 类、"登录页面"和"显示页面"组成。

（1）登录类 LoginBeans 包含 2 个属性和 5 个方法，其中 check()方法表示判断登录是否成功，当输入用户名称为"yang"且密码为"yang"时，登录才会成功。

```
public class LoginBean {

    private String loginName = null;
    private String password = null;

    public void LoginBean() {
    }

    public String getLoginName() {
        return loginName;
    }

    public String getPassword() {
        return password;
    }

    public void setLoginName(String loginName) {
        this.loginName = loginName;
    }

    public void setPassword(String password) {
        this.password = password;
```

```java
    }
    public boolean check(){
        if ("yang".equals(loginName) && "yang".equals(password)){
            return true;
        }else{
            return false;
        }
    }
}
```

（2）登录页面 login.jsp。

页面 login.jsp 提供了 2 个输入框和 2 个按钮，分别代表用户名称输入框、密码输入框、登录按钮和重置按钮，通过表单管理这些输入标签，具体页面实现代码如下：

```jsp
<%@ page contentType="text/html;charset=UTF-8" language="java" %>
<html>
<head>
    <title>Title</title>
</head>
<body bgcolor="cyan">

<form method="post" action="show.jsp" name="form1">

    <p>输入用户名称：
        <input type="text" name="loginName" size="20"/>

        <br>输入用户密码：
        <input type="password" name="password" size="20"/>

        <br>确认用户信息：   
        <input type="submit" name="submit" value="提交" size="6"/>
        <input type="reset" name="reset" value="重置" size="6"/>

</form>
</body>
</html>
```

上面代码中，当单击"登录"按钮后，页面跳转到 show.jsp 页面，实现登录结果显示。

（3）显示页面 show.jsp。

根据案例分析，显示页面 show.jsp 主要实现登录页面的信息。当登录成功后页面跳转到显示页面 show.jsp，登录失败后跳转到登录页面 login.jsp。

```jsp
<%@ page contentType="text/html;charset=UTF-8" language="java" %>
<%@ page import="mybean.maths.LoginBean" %>
<html>
<head>
    <title>Title</title>
</head>
<body bgcolor="cyan">

    <jsp:useBean id="stu" class="mybean.maths.LoginBean" scope="page">
```

```
</jsp:useBean>
            <jsp:setProperty name="stu" property="*"></jsp:setProperty>
            <%
                if (stu.check()){
            %>
            <h2>
                欢迎 <jsp:getProperty name="stu" property="loginName"></jsp:getProperty> 进入考生报名系统
            </h2>

            <%
            }else{
            %>
            <h2>
                登录失败 <jsp:forward page="login.jsp"></jsp:forward>
            </h2>

            <%
                }
            %>
            <p>您登录的信息是:
                <br>用户名称:
                <jsp:getProperty name="stu" property="loginName"></jsp:getProperty>
                <br>用户密码:
                <jsp:getProperty name="stu" property="password"></jsp:getProperty>

        </body>
        </html>
```

运行结果

启动运行案例中的 login.jsp 页面，运行结果如图 4-2 所示。

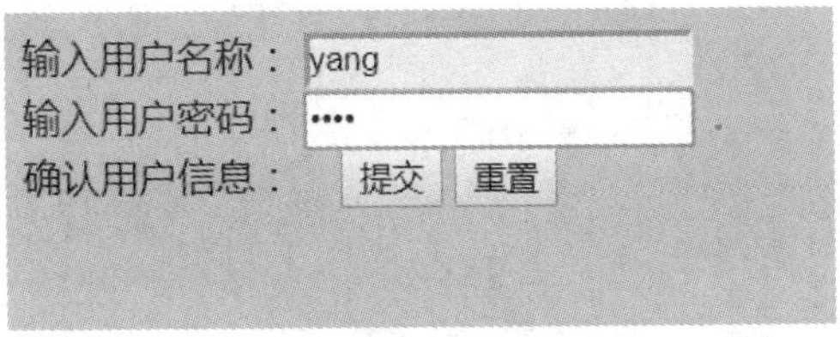

图 4-2 登录页面

输入正确的用户名称和密码后跳转到 show.jsp 页面，运行结果如图 4-3 所示。

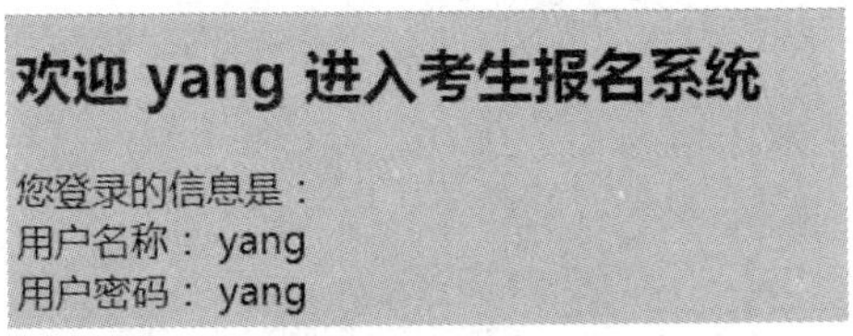

图 4-3 显示页面

输入错误的用户名称或密码后，跳转到 login.jsp 页面，运行结果如图 4-4 所示。

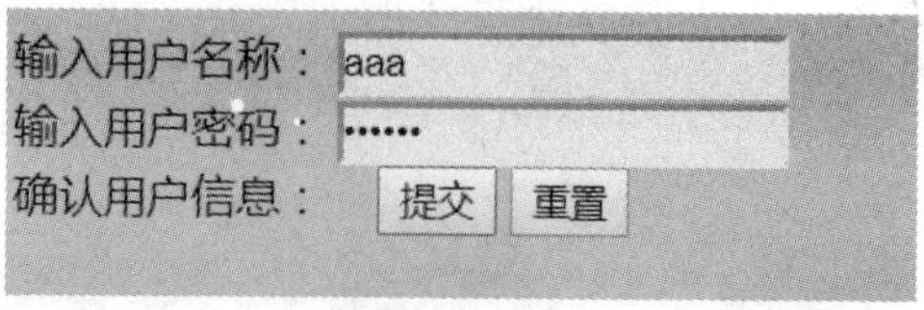

图 4-4　用户名称或密码输入错误的跳转页面

4.4　Excel 解析收到的信息

在 JSP 文件中设计表格后，运行页面，用 Excel 打开 JSP 文件，掌握 page 指令的用法。

案例描述

掌握 page 指令的用法。案例要求如下：
- 在 JSP 页面中设计表格。
- 下载 JSP 文件后用 Excel 打开 JSP 文件。

案例分析

根据案例描述，采用 JSP 技术实现该案例。

案例实现

根据案例分析，设计的页面代码如下：

```
<%@ page contentType="application/vnd.ms-excel;charset=UTF-8" language="java" %>
<html>
<head>
    <title>用Excel解析收到的信息</title>
</head>
<body bgcolor="cyan">

    <font size="4" color="blue">

        <table width="350" bordercolor="#FF0000" height="290" border="10"align="center" cellpadding="4" cellspacing="2">
            <tr align="center" height="30" valign="middle"bordercolor="#0000FF">
                <th>编号</th>
                <th>姓名</th>
                <th>性别</th>
                <th>工作单位</th>
            </tr>

            <tr align="center" valign="top" bordercolordark="#00FF00"
```

```
bordercolorlight="#FF0000">
                <th>01</th>
                <th>杨得力</th>
                <th>男</th>
                <th>河北XX大学</th>
            </tr>

            <tr align="right" valign="baseline" bgcolor="#CC99FF">
                <th>02</th>
                <th>周国正</th>
                <th>女</th>
                <th>北京XX大学</th>
            </tr>

        </table>

    </font>

</body>
</html>
```

运行结果

运行案例中的项目，运行结果如图 4-5 所示。

图 4-5　Excel 解析收到的信息

习题

1. 请简单叙述 include 指令标记和 include 动作标记的不同。
2. 是否允许一个 JSP 页面为 contentType 设置 2 次不同的值。
3. 简述 page 指令和 include 指令的作用。
4. JSP 常用的基本动作有哪些？简述其作用。
5. 编写 2 个 JSP 页面：main.jsp 页面和 lader.jsp 页面，将 2 个 JSP 页面保存在同一 Web 服务目录中。main.jsp 页面使用 include 动作标记动态加载 lader.jsp 页面。lader.jsp 页面可以计算并显示梯形的面积。当 lader.jsp 页面被加载时，获取 main.jsp 页面中 include 动作标记的 param 子标记提供的梯形的上底、下底和高的值。

第 5 章 内置对象技术

内置对象技术在 JSP 编程中比较重要，JSP 中涉及 request、response、out、application、session、page、pageContext、config 和 exception 共 9 个内置对象，内置对象是指无须程序员创建，可直接通过它的名称调用的对象。不同的内置对象，功能和作用不一样，本章通过项目案例的方式讲解这些内置对象的使用规则、作用及在项目中的应用。

本章任务

（1）内置对象的种类及作用；
（2）内置对象的常用功能；
（3）内置对象项目应用。

重点内容

（1）掌握对象的种类及作用；
（2）掌握内置对象的常用功能；
（3）掌握内置对象项目应用。

难点内容

内置对象项目应用。

5.1 内置对象概述

JSP 中的内置对象是指无须程序员创建，可直接在 JSP 页面通过对象名使用的对象。实际上，这些对象也是需要创建的，只不过是不需要程序员手动创建，而是由 JSP 的 Web 容器实现创建的。

在 JSP 中，共有 request 客户端请求对象、out 服务器端输出对象、response 服务器端响应对象、session 会话对象、application 应用程序对象、page 页面对象、pageContext 页面上下文对象、config 服务器端配置参数对象、exception 异常对象等 9 个。其中，常用对象有 request、out、response、session 和 application 5 个；不常用内置对象有 page、pageContext、config 和 exception 异常对象 4 个。本章重点介绍 5 个常用内置对象，通过项目综合案例讲述它们的用途，对其他 4 个不常用内置对象进行概念阐述。

5.1.1 request 对象

request 客户端请求对象，属于 HttpServletRequest 接口类型，内置对象 request 已经实现该接口。客户端通过该对象将客户端的信息带给服务器，信息包括：客户端的头信息、提交的各种参数等。HttpServletRequest 类的描述见表 5-1。

表 5-1 HttpServletRequest 类的描述

类名称	HTTPServletRequest	中文名称	客户端请求类	类型	接口
类描述	提供获取客户端各种请求信息功能等				
序号	方法名	参数	返回类型	功能说明	
1	getParameter	Strings：参数名	String	根据参数名 s 获取对应的参数值。例如客户端传值到服务器端	
2	getContextPath	无	String	返回当前项目的路径	
3	getQueryString	无	String	获取客户端 get 请求地址栏参数	
4	getCookies	无	Cookies[]	获取当前客户端当前应用程序所有的 Cookies 对象	
5	getServletPath	无	String	获取当前客户端页面请求地址	
6	getAttribute	Strings：属性名称	Object	获取属性名称为 s 的值	
7	getSession	无	HttpSession	获取当前会话对象	
8	getHeader	String name：头文件名	String	获取 HTTP 协议定义的文件头信息	
9	getHeaders	String name：头文件信息名称	Enumeration<String>	返回指定名称 name 的所有值	
10	getHeaderNames	无	Enumeration<String>	返回所有 request Header 的名字	
11	getMethod	无	String	获取客户端向服务器端数据传递方式	
12	getProtocol	无	String	获取客户端向服务器端传送数据的协议名称	
13	getRequestURI	无	String	获取发出请求字符串的客户端地址，不包括请求的参数	
14	getRequestURL	无	String	获取发出请求字符串的客户端地址	
15	getRealPath	无	String	返回当前请求文件的绝对地址	
16	getRemoteAddr	无	String	获取客户端的 IP 地址	
17	getRemoteHost	无	String	获取客户端的主机名	
18	getRemotePort	无	String	返回客户端的端口号	
19	getServerName	无	String	获取服务器名称	
20	getServerPort	无	String	获取服务器的端口号	

Cookie 是用来存储在客户端上的应用程序数据，它们保存了大量轨迹信息。在 Servlet 技术基础上，JSP 提供对 HTTP Cookie 的支持。例如在记住密码的登录中就可以应用，用户在正确登录系统后，服务器将用户的身份信息写入 Cookie 对象中，下次登录时先向 Cookie 对象提取用户信息，如果提取到了，则按照提取的信息进行身份验证，这样就可以实现免密钥登录。

Cookie 数据存储时采用类似 key-value 键值对的形式。如：

```
Cookie cookie = new Cookie( "key" , "value" );
```

Cookie 类的描述见表 5-2。

表 5-2　Cookie 类的描述

类名称	Cookie	中文名称	客户端存储类	类　　型	实体类
类描述	提供获取客户端数据存取操作等				
序　号	方法名	参　　数	返回类型	功能说明	
1	setDomain	String pattern：域名匹配模式	void	设置 Cookie 的域名	
2	getDomain	无	String	获取 Cookie 的域名	
3	setMaxAge	int expiry：时间长度，单位为秒	void	设置 Cookie 有效期，以秒为单位，默认有效期为当前 Session 的存活时间	
4	getMaxAge	无	Int	获取 Cookie 有效期，以秒为单位，默认值为-1，表明 Cookie 会活到浏览器关闭为止	
5	getName	无	String	返回 Cookie 名称	
6	getValue	无	String	获取 Cookie 的值	
7	setPath	String uri	void	设置 Cookie 的路径，默认为当前页面目录下的所有 URL，含目录下的所有子目录	
8	getPath	无	String	获取 Cookie 的路径	
9	setSecure	boolean flag：是否加密传输	void	设置 Cookie 是否加密传输，true 表示加密，false 表示无效加密	
10	setComment	String memo：描述内容	void	设置 Cookie 描述内容	
11	getComment	无	String	获取 Cookie 描述内容	

使用 Cookie 的步骤如下：

- 创建 Cookie 对象，例如 Cookie cookie = new Cookie("key","value");
- 设置 Cookie 有效期，以秒为单位，如：cookie.setMaxAge(1000);
- 将 Cookie 发送至 HTTP 响应头中，调用 response.addCookie()方法。

5.1.2　out 内置对象

out 字符输出流内置对象，属于 JspWriter 类型，用于服务器端向客户端页面输出字符信息。JspWriter 类是从 Writer 类继承而来的，Writer 类有 write 输出方法，而子类 JspWriter 中有 print 输出方法，它们之间存在差异。

print 方法在子类 JspWriter 中定义，writer 方法在父类 Writer 中定义，重载的 print

方法可以将各种类型的数据转换成字符串的形式输出，而重载的 writer 系列方法中能输出字符、字符数组和字符串等与字符相关的数据。JspWriter 类型的 out 对象使用 print 方法和 write 方法都可以输出字符串，但是，如果字符串对象的值为 null，则 print 方法输出内容为"null"的字符串，而 write 方法则会抛出 NullPointerException 异常。

JspWriter 类常用的功能方法见表 5-3。

表 5-3 JspWriter 类常用的功能方法

类名称	JspWriter	中文名称		页面输出类	类　型	抽象类
类描述	提供向 JSP 页面输出字符信息等					
序　号	方法名	参　数		返回类型	功能说明	
1	newLine	无		void	换行，页面执行一个空格	
2	print	boolean var1：布尔型参数		void	向页面输出参数 var1	
3	print	char var1：字符参数		void	向页面输出参数 var1	
4	print	int var1：整型参数		void	向页面输出参数 var1	
5	print	long var1：长整型参数		void	向页面输出参数 var1	
6	print	float var1：单精度参数		void	向页面输出参数 var1	
7	print	double var1：双精度参数		void	向页面输出参数 var1	
8	print	char[] var1：字符数组参数		void	向页面输出参数 var1	
9	print	String var1：字符串参数		void	向页面输出参数 var1	
10	print	Object var1：对象参数		void	向页面输出参数 var1	
11	printIn	print 中各种类型参数		void	向页面输出内容后留一个空格	
12	write	int c：字符参数		void	向页面输出字符 c	
13	write	String str：字符串参数		void	向页面输出字符串 str	
14	write	char[] c：字符数组		void	向页面输出字符数组 c	
15	clear	无		void	清除缓冲区，没有任何内容向页面显示	
16	clearBuffer	无		void	如果先调用 flush 方法，再执行此方法，则会在页面显示缓冲区内容，再清空缓冲区	

关于 out 内置对象输出功能有几点需要注意：
- write 方法向页面输出仅限于字符、字符串；print 方法可以向页面输出各种类型的数据。
- newLine 方法与 printIn 方法在页面输出中不是换行，而是一个空格。

5.1.3 response 内置对象

response 内置对象是服务器向客户端做出响应的对象，是 HttpServletResponse 类的一个实例，它是一个接口类型，由 JSP 容器实现实例化。通过 response 设置 HTTP 的状态和向客户端发送数据，向客户端响应的数据格式类型有多种形式，通过调用 response 的 setHeader 方法实现，默认响应给客户端的格式为字符，也可以返回字节类型数据以实现文件发送给客户端。

response 响应的常用头信息包括的内容见表 5-4。

表 5-4 response 响应常用头信息

序号	响应头	描述
1	Allow	允许服务器支持的 request 的请求方法（GET，POST 等）
2	Cache-Control	设置响应文档安全缓存。通常取值为 public，private 或 no-cache 等。public 意味着文档可缓存，private 意味着文档只为单用户服务并且只能使用私有缓存。no-cache 意味着文档不被缓存
3	Connection	命令浏览器是否要使用持久的 HTTP 连接。close 值命令浏览器不使用持久 HTTP 连接，而 keep-alive 意味着使用持久化连接
4	Content-Disposition	设置浏览器要求用户将响应以给定的名称存储在磁盘中，适用于文件下载
5	Content-Encoding	设置传输页面的编码规则
6	Content-Language	文档使用的语言
7	Content-Length	响应字节的长度
8	Content-Type	文档的 MIME 类型
9	Expires	设置过期并从缓冲中移除
10	Last-Modified	设置文档最后的修改时间
11	Location	在 300 秒内，有一个状态码的响应地址，浏览器会自动重连然后检索新文档
12	Refresh	设置浏览器多久请求刷新一次页面
13	Set-Cookie	设置当前页面对应的 Cookie

HttpServletResponse 接口提供了许多功能，表 5-5 列出了项目中常用的一些功能。

表 5-5 HttpServletResponse 类

类名称	HttpServletResponse	中文名称		服务器响应类	类型	接口
类描述	服务器端向客户端提供操作功能					
序号	方法名	参数		返回类型	功能说明	
1	encodeRedirectURL	String url：重定向地址		String	对 sendRedirect()方法使用的 URL 进行编码	
2	encodeURL	String url：编码字符		String	将参数 URL 编码，回传包含 Session ID 的 URL	
3	containsHeader	String name：响应头名称		boolean	返回指定的 name 响应头是否存在，true 表示存在，false 表示不存在	
4	isCommitted	无		boolean	返回响应是否都已经提交到客户端，true 表示已提交完成，false 表示未提交完成	
5	addCookie	Cookie cookie：带给客户端 Cookie		void	添加指定 Cookie 到响应对象中，并将其带回客户端	
6	addDateHeader	String name：名称；Long date：日期值		void	添加指定名称响应头和日期值	
7	addHeader	String name：名称；String value：值		void	添加指定响应头名称和对应的值	
8	flushBuffer	无		void	将任何缓存中的内容写入客户端	
9	reset	无		void	清除缓存中的所有数据，包括状态码和各种响应头	

续表

序号	方法名	参数	返回类型	功能说明
10	resetBuffer	无	void	清除基本的缓存数据，不包括响应头和状态码
11	sendError	int sc：状态码	void	使用 sc 状态码向客户端发送一个出错响应，然后清除缓存
12	sendError	int sc：状态码；String msg：消息	void	使用 sc 状态码和 msg 消息向客户端发送一个出错响应
13	sendRedirect	String location：地址	void	使用指定的 location 向客户端发送一个临时的间接响应
14	setBufferSize	int size：缓冲大小	void	设置响应体的缓存区大小
15	setCharacterEncoding	String charset：编码名称	void	指定响应的编码集（MIME 字符集），例如 UTF-8
16	setContentType	String type：类型	void	设置响应的内容类型
17	getOutputStream	无	OutputStreamWriter	获取字节输出流对象

5.1.4 session 内置对象

session 内置对象是 HttpSession 接口的实例对象，JSP Web 容器实现了 HttpSession 接口，并创建了 session 内置对象。每一次客户端访问服务器时都会建立一次会话连接，创建一个 session 对象。session 的会话 ID 是唯一的，默认情况下，它允许会话跟踪，利用跟踪可以实现不同 JSP 页面值的传递，将需要传递的值写入 session 中，其他页面使用时再从 session 中获取，例如限制未登录的用户不允许访问页面；当然，如果不允许会话跟踪，就无法实现页面间通过 session 传值；如果要禁止 session 跟踪，只要设置 JSP 页面编译指令 page 的属性值 session="false"。

session 内置对象可以通过 setAttribute()方法来存储值，通过 getAttribute()方法获取存储的数值。session 中的值在其会话期间内不会丢失，只有会话期结束才会消失。具体的操作参照 HttpSession 接口中关于这些方法的定义。

HttpSession 接口提供了会话操作功能，常用的功能请参考表 5-6：

表 5-6 HttpSession 类

类名称	HttpSession	中文名称	服务器响应类	类型	接口
类描述	提供会话操作功能				
序号	方法名	参数	返回类型	功能说明	
1	getAttribute	String name：属性名	Object	根据属性名 name 获取属性值	
2	getAttributeNames	无	Enumeration	获取 session 中所有的属性名称	
3	getCreationTime	无	long	返回 session 对象被创建的时间，以毫秒为单位，从 1970 年 1 月 1 号凌晨开始算起	
4	getId	无	String	获取 session 的 id。	
5	getLastAccessedTime	无	long	返回客户端最后访问的时间，以毫秒为单位，从 1970 年 1 月 1 号凌晨开始算起	

续表

序号	方法名	参数	返回类型	功能说明
6	getMaxInactiveInterval	无	Int	返回最大时间间隔，以秒为单位，Servlet 容器将会在这段时间内保持会话打开
7	invalidate	无	void	将 session 无效化，解绑任何与该 session 绑定的对象
8	isNew	无	void	判断 session 是否为一个新的客户端，或者客户端是否拒绝加入 session
9	removeAttribute	String name：属性名称	void	删除指定属性名 name 的属性对象
10	setAttribute	String name：属性名称；Object value：属性值	void	根据指定的属性名称 name 设置属性值 value 并绑定到 session 中
11	setMaxInactiveInterval	Int interval：间隔时间，单位为秒	void	设置会话保持连接的时间

5.1.5 application 内置对象

application 内置对象属于 javax.Servlet.ServletContext 的实例，而 ServletContext 是接口，由 JSP 容器实现该接口。application 内置对象代表 JSP 所属的 Web 应用本身，可以存储应用程序级别的变量，应用于 JSP 页面或者 Servlet 中，只要应用程序处于运行中，这些存储的变量值将一直存在；如果应用程序结束运行，变量中的值就会消失。ServletContext 接口常用的方法见表 5-7。

表 5-7 ServletContext 接口常用的方法

类名称	ServletContext	中文名称	Servlet 上下文	类型	接口
类描述	提供 Servlet 上下文操作功能				
序号	方法名	参数	返回类型	功能说明	
1	getAttribute	String name：属性名	Object	根据属性名 name 获取属性值	
2	setAttribute	String name：属性名；String value：属性值	void	设置属性名称 name 的属性值为 value	
3	removeAttribute	String name：属性名	void	移除指定名字 name 的 Servlet 容器变量	
4	getInitParameter	String param：参数名称	String	根据参数 param 获取参数值	
5	getContext	String var1：地址	String	返回一个指定 URL 地址的 ServletContext 对象	
6	getMimeType	String var1：文件地址	String	返回指定文件的文件类型，如果文件类型未知，则返回 null。文件类型由 Servlet 容器的配置决定并在一个 web-app 中被指定。一般情况下的文件类型是："text/html" 或 "image/gif"	

续表

序号	方法名	参数	返回类型	功能说明
7	getResourcePaths	String url：地址	Set<String>	返回一个存储 web-app 中所有资源路径的 Set（集合）。路径以"/"结尾表示一个子目录，并以"/"开头表示一个对于 web-app 的相对路径
8	getResource	String path：资源文件路径	URL	返回由 path 指定的资源路径对应的一个 URL 对象，该 path 必须以"/"开头并作为当前目录的相对位置
9	getResourceAsStream	String path：资源文件路径	InputStream	返回一个由 String path 指定位置资源的 InputStream。返回的 InputStream 可以是任意类型和长度
10	getRealPath	String path：虚拟路径	String	返回一个字符串，指定虚拟路径对应的真实路径（完整路径）
11	getInitParameterNames	无	Enumeration	返回上下文定义的所有参数名称枚举，如果为空则返回空枚举函数
12	addServlet	String var1：Serlvet 名称；String var2：类名称	void	以编程方式声明一个 Servlet。它添加 Servlet 名称和 Class 类名称到 Servlet 上下文对象
13	addServlet	String var1：Serlvet 名称；Servlet var2：Servlet 对象	void	以编程方式声明一个 Servlet。它添加 Servlet 名称和 Servlet 实例到 Servlet 上下文
14	addServlet	String var1：Servlet 名称；Class<? extends Servlet> var2：Servlet 对象	void	以编程方式声明一个 Servlet。它添加 Servlet 名称和 Servlet 类的一个实例到 Servlet 上下文
15	createServlet	Class<T> var1	T	根据参数 T 类型实例化 Servlet 对象
16	addFilter	String var1：过滤器名称；String var2：过滤器类名称	void	以编程方式声明一个 Filter。它添加过滤器的名称和 Class 名称到 Web 应用
17	addFilter	String var1：过滤器名称；Filter var2：过滤器对象	void	以编程方式声明一个 Filter。它添加过滤器名称和 Filter 实例到 Web 应用
18	addFilter	String var1：过滤器名称；Class<? extends Filter> var2：过滤器对象	void	以编程方式声明一个 Filter。它添加过滤器名称和 Filter 实例到 Web 应用
19	createFilter	Class<T> var1：过滤器对象	T	根据 T 类型实例化一个过滤器对象
20	addListener	T var1：监听器对象	void	向 ServletContext 添加指定 Class 名称的监听器

续表

序号	方法名	参数	返回类型	功能说明
21	addListener	String var1：监听器名称	void	向 ServletContext 添加一个给定的监听器
22	addListener	Class<? extends EventListener> var1：监听器对象	void	向 ServletContext 添加指定 class 类型的监听器
23	createListener	Class<T> var1：监听器对象	T	根据 T 类型实例化一个监听器对象
24	getClassLoader	无	ClassLoader	获取类请求服务对象

5.1.6　page 内置对象

page 代表页面本身，通常没有太大的用处，也就是 Servlet 中的 this 对象，其类型是 Servlet。

5.1.7　pageContext 内置对象

pageContext 内置对象是 javax.Servlet.jsp.PageContext 的实例，该实例代表 JSP 页面的上下文，使用该实例可以访问页面中的共享数据。常用的方法有 getServletContext()和 getSevletConfig()等。

5.1.8　config 内置对象

config 内置对象属于 javax.Servlet.ServletConfig 的实例，该实例代表 JSP 的配置信息，常用的方法有 getInitparameter(String paramName)及 getInitParametername()等方法。事实上，JSP 页面通常无需配置，也就不存在配置信息。因此该对象更多地在 Servlet 中有效。

5.1.9　exception 内置对象

exception 内置对象是 java.lang.Throwable 的实例，该实例代表其他页面中的错误和异常。只有当页面发生错误时，并且编译指令 isErrorPage 属性为 true，该对象才可以使用。常用的方法有 getMessage()和 printStackTrace()等。

5.2　内置对象的使用

内置对象在 Java Web 开发中经常将多个对象组合在一起发挥作用，本节通过简单案例的形式分别对 request、out、response、session 和 application 等 5 个内置对象的常用功能作项目应用阐述。

5.2.1 手机信息采集

手机信息采集案例展示内置对象 request 和 out 常用功能的使用。

案例描述

设计手机信息采集页面和展示采集到的手机信息显示页面，通过 request 和 out 内置对象实现其功能，手机信息采集的内容包括：手机编号、手机品牌、存储、分辨率、屏幕尺寸、操作系统、电池容量和 CPU 等信息。另一个信息显示来自客户端的信息以及已经采集到的手机信息，客户端信息包括：操作系统、浏览器、浏览器版本、IP 地址、计算机名称、端口号、访问协议、会话 ID、请求地址等信息。

案例分析

根据案例描述，需要有采集手机信息的页面和显示采集到的手机信息及客户端信息页面，因此需要设计两个页面，phone.jsp 页面实现采集手机信息，doPhone.jsp 页面显示手机及客户端信息。案例描述中提到了具体的手机信息内容，为了更好地管理这些信息，设计一个 JavaBean 手机 Phone 类来描述这些信息；要求采集客户端的操作系统、浏览器及浏览器版本信息，而这些信息在 request 内置对象的头信息 user-Agent 中，而要得到这些信息，需要设计一个 TerminateDevice 类。phone.jsp、doPhone.jsp、Phone.java 和 TerminateDevice 之间形成了如图 5-1 所示的关系。

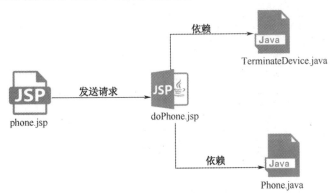

图 5-1 手机信息采集文件之间的关系

图 5-1 中，用户通过 phone.jsp 页面填写手机各种所需的信息，发送到页面 doPhone.jsp，该页面将接收到的信息进行处理，处理客户端信息使用了 TerminateDevice 类和 request 内置对象，而显示手机信息需要使用 Phone.java 对象。

案例实现

根据案例分析，手机信息采集案例需要使用 phone.jsp 和 doPhone.jsp 两个页面文件及 TerminateDevice 和 Phone 两个 Java 类文件；其中 phone.jsp 和 doPhone.jsp 页面通过编译指令 include 引入 init_bootstrap.jsp 页面，该页面配置了 BootStrap 框架。

1. init_bootstrap.jsp

init_bootstrap.jsp 页面做了 BootStrap 框架的配置，当需要使用该框架的页面时，直接在该页面中通过编译指令 include 引入它即可。具体参数配置如下：

```jsp
<%--
  Created by IntelliJ IDEA.
  User: wph-pc
  Date: 2017/8/1
  Time: 14:57
  To change this template use File | Settings | File Templates.
--%>
<%@ page contentType="text/html;charset=UTF-8" language="java" %>
<%
    String path = request.getContextPath();
    String basePath = request.getScheme()+"://"+request.getServerName()+":"+request.getServerPort()+path;
%>
<html>
<head>
    <title>JSP程序设计项目教程</title>
    <meta http-equiv="Content-Type" content="text/html; charset=utf-8"/>
    <meta http-equiv="X-UA-Compatible" content="IE=edge,chrome=1">
    <meta name="Author" content="kesun">
    <meta name="viewport" content="width=device-width, initial-scale=1, maximum-scale=1">

    <link href="<%= basePath %>/css/bootstrap.min.css" rel="stylesheet" />
    <link href="<%= basePath %>/css/bootstrapValidator.min.css" rel="stylesheet" />
    <link href="<%= basePath %>/plugins/bootstrap/bootstrap-treeview.min.css" rel="stylesheet" />
    <link href="<%= basePath %>/css/bootstrap-datetimepicker.min.css" rel="stylesheet" />
    <link href="<%= basePath %>/css/toastr.min.css" rel="stylesheet" />
    <script src="<%= basePath %>/script/common/jquery-1.10.2.min.js"></script>
    <script src="<%= basePath %>/plugins/bootstrap/bootstrap.min.js"></script>
    <script src="<%= basePath %>/plugins/bootstrap/bootstrapValidator.min.js"></script>
    <script src="<%= basePath %>/plugins/bootstrap/toastr.min.js"></script>
    <script src="<%= basePath %>/plugins/bootstrap/bootstrap-treeview.min.js"></script>
    <script src="<%= basePath %>/plugins/bootstrap/bootstrap-datetimepicker.min.js"></script>
    <script src="<%= basePath %>/plugins/bootstrap/bootstrap-datetimepicker.zh-CN.js"></script>
    <script src="<%= basePath %>/plugins/bootstrap/bootstrapMessageDialog.js"></script>

    <script src="<%= basePath %>/script/common/public.js"></script>
    <script src="<%= basePath %>/script/common/project.js"></script>
```

```html
            <script src="<%= basePath %>/script/common/mybootstrap.js">
</script>
            <script src="<%= basePath %>/script/common/bootstrapvalidators.js">
</script>

            <script>
                var kbValidator=new MyBootstrapValidator();
                $(function(){
                    toastr.options = {
                        positionClass: "toast-bottom-center",
                        onclick: null,
                        showDuration: "300",
                        hideDuration: "1000",
                        timeOut: "2000",
                        extendedTimeOut: "1000",
                        showEasing: "swing",
                        hideEasing: "linear",
                        showMethod: "fadeIn",
                        hideMethod: "fadeOut"
                    };
                });
            </script>
            <style>
                td {
                    background-color: #fff;
                    border: solid 10px #ffffff;
                    color: rgba(0, 0, 0, 0.99);
                    font-size: 12px;
                    font-family: Helvetica Neue,Helvetica,PingFang SC,Hiragino Sans GB,Microsoft YaHei,Noto Sans CJK SC,WenQuanYi Micro Hei,Arial,sans-serif;
                    margin: 0;
                    padding: 0;
                }
                .bs-callout {
                    padding: 20px;
                    margin: 20px 0;
                    border: 1px solid #eee;
                    border-left-width: 1px;
                    border-left-color: rgb(238, 238, 238);
                    border-left-width: 5px;
                    border-radius: 3px;
                }
                .bs-callout-danger {
                    border-left-color: #ce4844;
                }
                .bs-callout-danger h4 {
                    color: #ce4844;
                }
                .bs-callout-warning {
                    border-left-color: #aa6708;
                }
```

```css
        .bs-callout-warning h4 {
            color: #aa6708;
        }
        .bs-callout-info {
            border-left-color: #1b809e;
        }
        .bs-callout-info h4 {
            color: #1b809e;
        }
        .mask {
            position: absolute; top: 0px; filter: alpha(opacity=60); background-color: #777;
            z-index: 1002; left: 0px;
            opacity:0.5; -moz-opacity:0.5;
            width: 100%;
            display:none;
            vertical-align: middle;
            text-align: center;
        }
    </style>
</head>
<body>
<div id="mask" class="mask">
    <img src="<%=basePath%>/images/loading.gif"/>
</div>
</body>
</html>
```

源代码中涉及的组件，读者可以根据配置从网络资源库自行下载或下载配套教材源代码。

2. phone.jsp 手机信息采集页面

phone.jsp 页面实现采集手机各种信息，页面没有做验证，采用了 BootStrap 框架，运行效果如图 5-2 所示。

图 5-2 手机信息采集页面效果图

```jsp
<%@ page import="java.io.IOException" %><%--
  Created by IntelliJ IDEA.
  User: wph-pc
  Date: 2018/10/26
  Time: 9:36
  To change this template use File | Settings | File Templates.
--%>
<%@ page contentType="text/html;charset=UTF-8" language="java" %>
<%@ include file="../../header/init_bootstrap.jsp"%>
<html>
<head>
    <title>request应用</title>
</head>
<body class="container" style="padding-top: 10px;">
    <div class="panel panel-primary">
        <div class="panel-heading">手机基本信息采集</div>
        <div class="panel-body">
            <form action="doPhone.jsp" method="post">
                <div class="form-group">
                    <label for="txtID">手机编号</label>
                    <input type="text"  class="form-control" name="id" placeholder="请输入手机编号" id="txtID"/>
                </div>
                <div class="form-group">
                    <label for="txtBrand">品牌</label>
                    <input type="text"  class="form-control" name="brand" placeholder="请输入手机品牌" id="txtBrand"/>
                </div>
                <div class="form-group">
                    <label for="txtStorage">机身存储</label>
                    <input type="text"  class="form-control" name="storage" placeholder="请输入手机存储, GB为单位" id="txtStorage"/>
                </div>
                <div class="form-group">
                    <label for="txtPPI">屏幕分辨率</label>
                    <input type="text"  class="form-control" name="ppi" placeholder="请输入手机屏幕分辨率" id="txtPPI"/>
                </div>
                <div class="form-group">
                    <label for="txtScreenSize">屏幕尺寸</label>
                    <input type="text"  class="form-control" name="screenSize" placeholder="请输入手机屏幕尺寸" id="txtScreenSize"/>
                </div>
                <div class="form-group">
                    <label for="txtOS">操作系统</label>
                    <input type="text"  class="form-control" name="os" placeholder="请输入手机操作系统" id="txtOS"/>
                </div>
                <div class="form-group">
                    <label for="txtBattery">电池容量</label>
                    <input type="text"  class="form-control" name="battery"
```

```html
                placeholder="请输入手机电池容量参数" id="txtBattery"/>
            </div>
            <div class="form-group">
                <label for="txtMemory">运行内存</label>
                <input type="text" class="form-control" name="memory"
placeholder="请输入手机运行的内存,以GB单位" id="txtMemory"/>
            </div>
            <div class="form-group">
                <label for="txtCPU">手机CPU</label>
                <input type="text" class="form-control" name="cpu"
placeholder="请输入手机运行的CPU参数,包括多核数据参数" id="txtCPU"/>
            </div>
            <button type="submit" class="btn btn-primary">提交</button>
        </form>
    </div>
</body>
</html>
```

3. TerminateDevice 类

TerminateDevice 类是用来根据客户端传递的"user-Agent"用户代理参数获取客户端的操作系统、浏览器及浏览器版本等信息，源代码如下：

```java
package chapter6.requestout;
import java.util.StringTokenizer;
import java.util.regex.Matcher;
import java.util.regex.Pattern;
/**
 * 客户端终端设备判断
 * Created by wph-pc on 2018/10/17.
 */
public class TerminateDevice {
    //浏览器用户代理信息
    private String userAgent="";
    public TerminateDevice(String ua)
    {
        this.userAgent=ua;
    }
    /*获取操作系统类型*/
    public String getOS()
    {
        if (userAgent.toLowerCase().indexOf("windows")>=0)
            return "windows";
        if (userAgent.toLowerCase().indexOf("mac")>=0)
            return "mac";
        if (userAgent.toLowerCase().indexOf("iphone")>=0)
            return "iphone";
        if (userAgent.toLowerCase().indexOf("ipad")>=0)
            return "ipad";
        if (userAgent.toLowerCase().indexOf("linux")>=0)
            return "linux";
        if (userAgent.toLowerCase().indexOf("android")>=0)
            return "android";
```

```
            if (userAgent.toLowerCase().indexOf("unix")>=0 || userAgent.
indexOf("sunos")>=0 || userAgent.indexOf("bsd")>=0)
                return "unix";
            return "";
        }
        /*获取浏览器信息*/
        private String getBrowserInfo()
        {
            String []temp=userAgent.split(" ");
            if (temp.length<3)
                return null;
            else
                return temp[temp.length-1];
        }
        /*获取浏览器类型*/
        public String getBrowser()
        {
            String browser=getBrowserInfo();
            if (browser==null)
                return null;
            else
            {
                String[] source=browser.split("/");
                if (source.length!=2)
                    return null;
                else
                    return source[0];
            }
        }
        /*获取浏览器版本号*/
        public String getBrowserVersion()
        {
            String browser=getBrowserInfo();
            if (browser==null)
                return null;
            else
            {
                String[] source=browser.split("/");
                if (source.length!=2)
                    return null;
                else
                    return source[1];
            }
        }
    }
```

4. Phone.java 类

Phone.java 类是手机信息 JavaBean 类，用来描述手机的基本信息。源代码如下：

```
package chapter6.requestout;

/**
```

```java
 * 手机基本信息类
 * Created by wph-pc on 2018/10/26.
 */
public class Phone {
    //手机编号
    private String id=null;
    //手机品牌
    private String brand=null;
    //存储
    private Integer storage=null;
    //分辨率
    private String ppi=null;
    //屏幕尺寸
    private String screenSize=null;
    //操作系统
    private String os=null;
    //运行内存
    private Integer memory=null;
    //电池容量
    private String battery=null;
    //cpu
    private String cpu=null;

    public String getId() {
        return id;
    }

    public void setId(String id) {
        this.id = id;
    }

    public String getBrand() {
        return brand;
    }

    public void setBrand(String brand) {
        this.brand = brand;
    }

    public Integer getStorage() {
        return storage;
    }

    public void setStorage(Integer storage) {
        this.storage = storage;
    }

    public String getPpi() {
        return ppi;
    }
```

```java
    public void setPpi(String ppi) {
        this.ppi = ppi;
    }

    public String getScreenSize() {
        return screenSize;
    }

    public void setScreenSize(String screenSize) {
        this.screenSize = screenSize;
    }

    public String getOs() {
        return os;
    }

    public void setOs(String os) {
        this.os = os;
    }

    public Integer getMemory() {
        return memory;
    }

    public void setMemory(Integer memory) {
        this.memory = memory;
    }

    public String getBattery() {
        return battery;
    }

    public void setBattery(String battery) {
        this.battery = battery;
    }

    public String getCpu() {
        return cpu;
    }

    public void setCpu(String cpu) {
        this.cpu = cpu;
    }
}
```

5. doPhone.jsp 页面

doPhone.jsp 页面实现将接收到的手机信息及客户端请求信息显示出来，具体实现源码如下：

```jsp
<%@ page import="chapter6.requestout.TerminateDevice" %>
<%@ page import="chapter6.requestout.Phone" %><%--
  Created by IntelliJ IDEA.
```

```jsp
  User: wph-pc
  Date: 2018/10/26
  Time: 10:59
  To change this template use File | Settings | File Templates.
--%>
<%@ page contentType="text/html;charset=UTF-8" language="java" %>
<%@ include file="../../header/init_bootstrap.jsp"%>
<html>
<head>
    <title>request与out综合应用</title>
</head>
<body class="container" style="padding-top: 20px;">
<div class="row">
    <div class="col-md-6 col-sm-12">
        <div class="panel panel-primary">
            <!-- Default panel contents -->
            <div class="panel-heading">客户端基本信息</div>
            <div class="panel-body">
                <p>request获取客户端信息应用，通过request可以判断客户端各种信息数据来源</p>
                <hr>
                <%
                    //获取客户端请求头信息
                    String header=request.getHeader("user-Agent");
                    //创建处理客户端操作系统、浏览器及浏览器类型对象
                    TerminateDevice terminate=new TerminateDevice(header);
                    StringBuilder sb=new StringBuilder();
                    //操作系统
                    sb.append("<dl><dt>操作系统</dt><dd>"+(terminate==null?"系统未识别":terminate.getOS())+"<dd></dl>");
                    //浏览器
                    sb.append("<dl><dt>浏览器</dt><dd>"+(terminate==null?"系统未识别":terminate.getBrowser())+"<dd></dl>");
                    //浏览器版本
                    sb.append("<dl><dt>浏览器版本</dt><dd>"+(terminate==null?"系统未识别":terminate.getBrowserVersion())+"<dd></dl>");
                    //IP地址
                    sb.append("<dl><dt>IP地址</dt><dd>"+request.getRemoteAddr()+"<dd></dl>");
                    //计算机名称
                    sb.append("<dl><dt>计算机名称</dt><dd>"+request.getRemoteHost()+"<dd></dl>");
                    //端口号
                    sb.append("<dl><dt>端口号</dt><dd>"+request.getRemotePort()+"<dd></dl>");
                    //访问协议
                    sb.append("<dl><dt>访问协议</dt><dd>"+request.getProtocol()+"<dd></dl>");
                    //session ID
                    sb.append("<dl><dt>session ID</dt><dd>"+request.getSession().getId()+"<dd></dl>");
```

```html
                    //请求地址
                    sb.append("<dl><dt>请求地址</dt><dd>"+request.getRequestURL()+"<dd></dl>");
                    out.print(sb.toString());
                %>
            </div>
        </div>
    </div>
    <div class="col-md-6 col-sm-12">
        <div class="panel panel-success">
            <!-- Default panel contents -->
            <div class="panel-heading">手机基本信息</div>
            <div class="panel-body">
                <p>request的getParameter方法应用及out输出应用</p>
                <hr>
                <%
                    //解决中文乱码
                    request.setCharacterEncoding("utf-8");
                    /*获取手机信息*/
                    Phone phone=new Phone();
                    phone.setId(request.getParameter("id")==null?"无数据":request.getParameter("id"));
                    phone.setBrand(request.getParameter("brand")==null?"无数据":request.getParameter("brand"));
                    phone.setStorage(request.getParameter("storage")==null?0:Integer.valueOf(request.getParameter("storage")));
                    phone.setPpi(request.getParameter("ppi")==null?"无数据":request.getParameter("ppi"));
                    phone.setStorage(request.getParameter("screenSize")==null?0:Integer.valueOf(request.getParameter("screenSize")));
                    phone.setOs(request.getParameter("os")==null?"无数据":request.getParameter("os"));
                    phone.setMemory(request.getParameter("memory")==null?0:Integer.valueOf(request.getParameter("memory")));
                    phone.setBattery(request.getParameter("battery")==null?"无数据":request.getParameter("battery"));
                    phone.setCpu(request.getParameter("cpu")==null?"无数据":request.getParameter("cpu"));
                    StringBuilder sbPhone=new StringBuilder();
                    //手机编号
                    sbPhone.append("<dl><dt>手机编号</dt><dd>"+phone.getId()+"<dd></dl>");
                    //手机品牌
                    sbPhone.append("<dl><dt>手机品牌</dt><dd>"+phone.getBrand()+"<dd></dl>");
                    //存储
                    sbPhone.append("<dl><dt>存储</dt><dd>"+phone.getStorage()+"<dd></dl>");
                    //分辨率
                    sbPhone.append("<dl><dt>分辨率</dt><dd>"+phone.
```

```
getPpi()+"<dd></dl>");
                        //屏幕尺寸
                        sbPhone.append("<dl><dt>屏幕尺寸</dt><dd>"+phone.
getScreenSize()+"<dd></dl>");
                        //操作系统
                        sbPhone.append("<dl><dt>操作系统</dt><dd>"+phone.
getOs()+"<dd></dl>");
                        //运行内存
                        sbPhone.append("<dl><dt>运行内存</dt><dd>"+phone.
getMemory()+"<dd></dl>");
                        //电池容量
                        sbPhone.append("<dl><dt>电池容量</dt><dd>"+phone.
getBattery()+"<dd></dl>");
                        //CPU
                    sbPhone.append("<dl><dt>CPU</dt><dd>"+phone.getCpu()+
"<dd></dl>");
                        out.print(sbPhone.toString());
                    %>
                </div>
            </div>
        </div>
    </body>
</html>
```

运行结果

运行案例中的 phone.jsp 页面，输入相应的手机参数信息，可得到如图 5-3 和图 5-4 所示的运行结果，当然，采集的信息不同，结果就不一样。

图 5-3 客户端信息

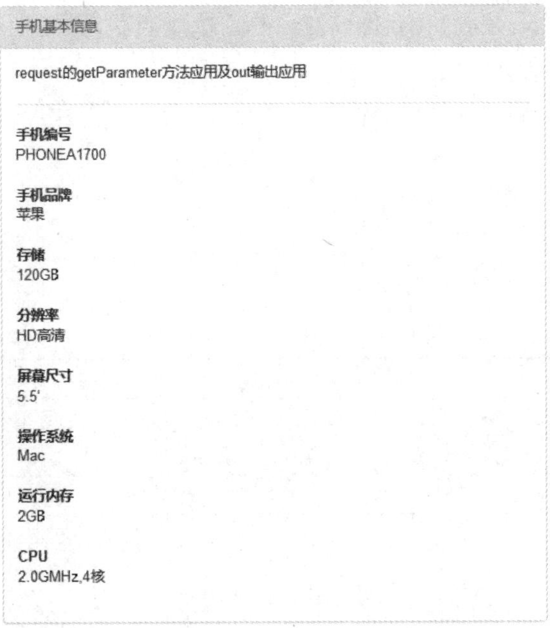

图 5-4　获取到客户端提交手机信息

5.2.2　验证码

验证码在项目中应用在很多地方，例如：身份验证需要提交验证码，信息发送或评论需要提交验证码等。验证码的作用是为了提升项目的安全，防止恶意的输入攻击。

案例描述

利用 JSP 技术，在页面上随机产生 10 位以内的数字验证码，并对验证码的数字进行着色，如图 5-5 所示。

图 5-5　用户登录验证码应用

在图 5-5 所示的登录页面中，验证码以图片的形式出现在页面上。

案例分析

根据案例描述要求，需要解决以下问题：

- 验证码的产生，控制在 10 位以内，并且是数字形式；
- 验证码如何着色；
- 验证码怎样以图片方式输出；
- 验证码如何显示在页面上。

按照项目待解决的问题，设计 Code 类，实现产生 10 位以内的随机数、随机颜色设置和根据随机数转换的图像等功能。此外，还需要设计一个 codeimage.jsp 页面，实现显示随机数验证码。Code 类设计见表 5-8。

表 5-8 Code 类

类 名 称	Code	中文名称	验证码	类 型	实体类
类 描 述	提供验证码创建、颜色设置及图像转换操作功能				
序 号	方法名	参 数	返回类型	功能说明	
1	getRandColor	无	Color	获取随机颜色	
2	getRand	int digit：随机数位数	String	获取 digit 位随机数字字符，digit 不能大于 10 或小于等于 0	
3	codeToImage	int width：图像宽度；int height：图像高度；String sRand：待转换图像的验证码	BufferedImage	根据图像宽度、高度及 sRand 验证码获取图像	

codeimage.jsp 验证码呈现程序流程如图 5-6 所示。

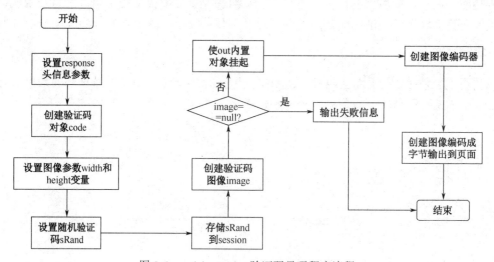

图 5-6 codeimage.jsp 验证码呈现程序流程

codeimage.jsp 不能直接显示验证码，如果直接调用，显示的是乱码字符，需要在另外的页面中设置 HTML 标签 img，通过其 src 属性设置为 codeimage.jsp 即可。

案例实现

案例实现由 Code 类和 imagecode.jsp 页面两个文件组成，Code 类实现验证码创建、着色和转换成图片；imagecode.jsp 实现页面显示。

1. Code 类

```java
package chapter6.response;

import java.awt.*;
import java.awt.image.BufferedImage;
import java.util.Random;

/**
 * 验证码操作
 * Created by wph-pc on 2018/10/10.
 */
public class Code {
    /*产生随机颜色*/
    public Color getRandColor() {//给定范围获得随机颜色
        Random random = new Random();
        int r = random.nextInt(255);
        int g = random.nextInt(255);
        int b = random.nextInt(255);
        return new Color(r, g, b);
    }
    /*产生digit位随机数*/
    public String getRand(int digit)
    {
        if (digit<=0 || digit >10) return "";
        StringBuilder sb=new StringBuilder();//字符串构建对象
        for(int i=0;i<digit;i++)
        {
            Random rand=new Random();
            sb.append(rand.nextInt(9));
        }
        return sb.toString();
    }
    /*根据参数width宽度和height高度创建sRand随机数字字符内存图像对象
    * @param width:图像宽度
    * @param height:图像高度
    * @param sRand:随机验证码,数字字符
    * @return 返回根据随机验证码的内存图像
    * */
    public BufferedImage codeToImage(int width, int height,String sRand)
    {
        BufferedImage image = new BufferedImage(width, height,
                BufferedImage.TYPE_INT_RGB);
        // 获取图形上下文
        Graphics g = image.getGraphics();
        //创建验证码对象
        Code code=new Code();
        g.fillRect(0, 0, width, height);

        //设定字体
```

```
            g.setFont(new Font("Times New Roman", Font.PLAIN, 18));
            //设置文字颜色
            g.setColor(code.getRandColor());
            g.drawString(sRand,5,15);
            // 图像生效
            g.dispose();
            return image;
        }
        public static void main(String[] args)
        {
            Code code=new Code();
            String rand=code.getRand(10);
            System.out.println(rand);
        }
    }
```

2. imagecode.jsp 页面

```
<%@ page import="java.awt.image.BufferedImage" %>
<%@ page import="java.awt.*" %>
<%@ page import="chapter6.response.Code" %>
<%@ page import="com.sun.image.codec.jpeg.JPEGImageEncoder" %>
<%@ page import="com.sun.image.codec.jpeg.JPEGCodec" %>
<%@ page import="java.io.IOException" %><%--
  Created by IntelliJ IDEA.
  User: wph-pc
  Date: 2018/10/25
  Time: 8:12
  To change this template use File | Settings | File Templates.
--%>
<%@ page contentType="text/html;charset=UTF-8" language="java" %>
<html>
<head>
    <title>验证码生成</title>
</head>
<body>
<%
    //设置页面不缓存
    response.setHeader("Pragma", "No-cache");
    response.setHeader("Cache-Control", "no-cache");
    response.setDateHeader("Expires", 0);

    //创建验证码对象
    Code code=new Code();
    // 在内存中创建图像
    int width = 60, height = 20;
    // 取随机产生的验证码(6位数字)
    String sRand =code.getRand(6);
    //验证码写入session对象,供验证使用
    session.setAttribute("code",sRand);
    //图像生效
```

```
            BufferedImage image=code.codeToImage(width,height,sRand);
            if(image==null)
            {
                out.print("系统验证码转换图像失败,原因可能是没有有效的验证码!");
                return;
            }
            /*以下两句非常重要,解决out内置对象与response.getOutputStream冲突问题*/
            out.clear();
            out = pageContext.pushBody();
            // 输出图像到页面
            try {
                JPEGImageEncoder encoder = JPEGCodec.createJPEGEncoder(response.getOutputStream());
                encoder.encode(image);
            } catch (IOException e) {
                e.printStackTrace();
            }
        %>
    </body>
</html>
```

运行结果

启动含有 img 标签的页面,并且其 src 设置为 codeimage.jsp,就可以看到验证码,验证码的字符值存放在 session 的内置对象 code 属性中。

5.2.3 用户身份验证

用户身份验证在项目中几乎是必不可少的模块,本节案例详细讲解如何利用内置对象技术实现用户身份验证。

案例描述

身份验证在软件项目中经常使用,要求利用内置对象技术实现用户身份登录,具体要求:

- 用户身份验证需要提供账号、密码及验证码;
- 须提供记住密码功能;
- 登录成功后跳转至主页面,主页面可实现历史访问人数统计;
- 如果用户没有合法登录,则无法直接跳转到主页面。

案例分析

根据案例描述,实现描述中的各种功能需要有登录页面模块 login.jsp、身份验证及记住密码处理功能页面 checkLogin.jsp、跳转使用的主页面 index.jsp,它们之间形成了图 5-7 的关系。

系统首先通过 login.jsp 页面获取客户端中的 Cookie,查看是否有保存的账号和密码,如果有,自动填写在账号与密码输入框中;如果没有,用户则手动输入;除获取账号和密码外，login.jsp 页面还要实现验证码信息的采集，可直接调用验证码案例的页面；验证码、账号与密码验证都在 checkLogin.jsp 页面中实现，如果验证不能通过，则跳转到 login.jsp 页面，否则将登录信息写入 Session 中保存；如果要记住密码，则将账号与密码写入 Cookie 中，最后跳转到 index.jsp 主页面，由 index.jsp 页面实现历史人数统计。login.jsp 页面、checkLogin.jsp 页面与 index.jsp 页面之间的关系如图 5-8 所示。

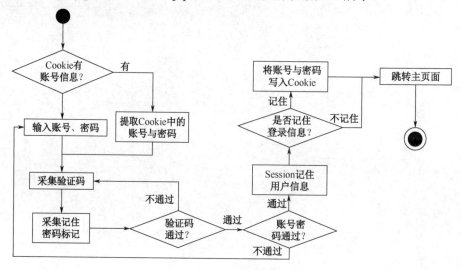

图 5-7 用户登录业务活动图

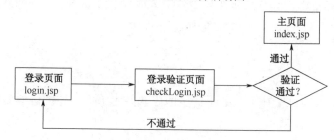

图 5-8 身份验证各页面之间的关系

图 5-8 中，登录验证页面 checkLogin.jsp 是核心功能，所有验证工作都在此页面完成，用户登录页面 login.jsp 实现采集用户信息及验证码，采集完成后提交给登录验证页面 checkLogin.jsp 处理，通过验证后，跳转至主页面 index.jsp。

在用户登录页面中，自定义了 JavaBean 用户类 User，用来表示用户对象信息，具体定义见表 5-9。

表 5-9 User 类

类 名 称	User	中文名称	用户类	类 型	实体类
类 描 述	提供用户属性表述				
成员变量					
序 号	变量名	访问权限	类 型	说 明	
1	number	private	String	用户账号	
2	nickName	private	String	用户昵称	
3	Password	private	String	登录密码	

续表

方法				
序号	方法名	参数	返回类型	功能说明
1	getNumber	无	String	获取用户账号
2	setNumber	String number：登录账号	void	设置用户账号
3	getNickName	无	String	获取昵称
4	setNickName	String nickName：昵称	void	设置昵称
5	getPassword	无	String	获取密码
6	setPassword	String password：密码	void	设置密码

案例实现

1. 登录页面 login.jsp

```jsp
<%--
  Created by IntelliJ IDEA.
  User: wph-pc
  Date: 2018/10/22
  Time: 22:39
  To change this template use File | Settings | File Templates.
--%>
<%@ page contentType="text/html;charset=UTF-8" language="java" %>
<%@include file="../../header/init_bootstrap.jsp"%>
<html>
<head>
    <title>用户登录记录密码</title>
</head>
<body class="container">
<%
    //定义存放账号变量number
    String number = "";
    //定义存放用户密码password变量
    String password = "";
    //获取客户端Cookie
    Cookie[] c = request.getCookies();
    /*找出是否存在账号和密码，如果存在则赋值给变量number和password*/
    if (c != null) {
        for (int i = 0; i < c.length; i++) {
            if ("number".equals(c[i].getName())) {
                number = c[i].getValue();
            } else if ("password".equals(c[i].getName())) {
                password = c[i].getValue();
            }
        }
    } else {
        number = " ";
        password = " ";
    }
%>
<div class="jumbotron">
```

```html
            <h1>用户身份验证综合案例</h1>
            <p>
                案例描述：用户身份验证，如果用户登录过，系统记住登录账号与密码。如果用户没有登录过，首次则需要输入正确的账号与密码。第二次登录时则无须输入账号与密码。另外还需要统计历史访问人数。
            </p>
            <p><a class="btn btn-primary btn-lg" href="#" role="button">Learn more</a></p>
        </div>
        <div class="row">
            <div class="col-md-12 col-sm-12 col-lg-12 col-xs-12">
                <form action="loginCheck.jsp" method="post">
                    <div class="form-group">
                        <label for="txtAccount">账号</label>
                        <input type="text" class="form-control" name="number" value="<%=number%>" placeholder="请输入账号" id="txtAccount"/>
                    </div>
                    <div class="form-group">
                        <label for="txtPwd">密码</label>
                        <input type="password" class="form-control" name="password" value="<%=password%>" id="txtPwd"/>
                    </div>
                    <div class="form-group">
                        <label for="txtCode">验证码</label>
                        <input type="text" class="form-control" name="code" maxlength="6" id="txtCode"/>
                        <img src="../response/codeimage.jsp" width="60" height="20" id="imgCode" onclick="refreshImg()">
                    </div>
                    <div class="checkbox">
                        <label>
                            <input type="checkbox" name="passcookies" id="ckRemember"/>记住密码
                        </label>

                    </div>

                    <button type="submit" class="btn btn-primary">登录</button>

                </form>
            </div>
        </div>

    <script>
        function refreshImg() {
            document.getElementById("imgCode").src="codeimage.jsp?timestamp="+(new Date()).getTime();
        }
    </script>
    </body>
</html>
```

2. 登录验证页面 checkLogin.jsp

```jsp
<%@ page import="business.User" %><%--
  Created by IntelliJ IDEA.
  User: wph-pc
  Date: 2018/10/22
  Time: 22:45
  To change this template use File | Settings | File Templates.
--%>
<%@ page contentType="text/html;charset=UTF-8" language="java" %>
<%@include file="../../header/init_bootstrap.jsp"%>
<html>
<head>
    <title>登录验证</title>
</head>
<body>
<%!
  /*页面上显示错误信息errMsg*/
  String showErrorTips(String errMsg)
  {
      //错误提示信息
      return "<div class='well'>"+errMsg+",请重新<a href='login.jsp'>登录</a></div>";
  }
%>
<%
    //设置客户端字符编码
    request.setCharacterEncoding("utf-8");
    //创建用户对象
    User user=new User();
    /*获取客户端账号、密码及是否保存密码标记*/
    user.setNumber(request.getParameter("number"));
    user.setPassword(request.getParameter("password"));
    String remember = request.getParameter("passcookies");
    /*验证码验证*/
    if (session.getAttribute("code")==null ||
        session.getAttribute("code") instanceof String ==false)
    {
        out.print(showErrorTips("系统没有获取到验证码信息"));
        return;
    }
    else
    {
        String code=session.getAttribute("code").toString();
        if (!code.equals(request.getParameter("code")))
        {
            out.print(showErrorTips("验证码错误"));
            return;
        }
    }
    /*验证账号与密码*/
```

```
            if (!"admin".equals(user.getNumber()) || !"admin".equals(user.getPassword()))  {
                out.print("账号"+user.getNumber()+";密码: "+user.getPassword());
                out.print(showErrorTips("账号或密码错误"));
                return;
            } else {
                //将用户信息写入Session中
                session.setAttribute("user",user);
                if (remember != null) {
                    Cookie c1 = new Cookie("number", user.getNumber());
                    Cookie c2 = new Cookie("password", user.getPassword());
                    //设置加密传输
                    c2.setSecure(true);
                    /*设置Cookie的时长为15天,以秒为单位*/
                    c1.setMaxAge(15*24*60*60);
                    c2.setMaxAge(15*24*60*60);
                    /*将设置的CooKie添加到response带回到客户端*/
                    response.addCookie(c1);
                    response.addCookie(c2);

                    out.println("<script>toastr.info('登录成功,并记住了账号与密码');</script>");
                } else {
                    out.println("<script>toastr.info('登录成功,但没有记住账号与密码');</script>");
                }
                //跳转主页面index.jsp
                response.sendRedirect("index.jsp");
            }
        %>

    </body>
    </html>
```

3. 主页面 index.jsp

```
<%--
  Created by IntelliJ IDEA.
  User: wph-pc
  Date: 2018/10/28
  Time: 16:45
  To change this template use File | Settings | File Templates.
--%>
<%@ page contentType="text/html;charset=UTF-8" language="java" %>
<%@ include file="../../header/init_bootstrap.jsp"%>
<html>
<head>
    <title>系统模拟主页面</title>
</head>
<body>
    <%
```

```
            //判断用户是否登录,未登录则直接跳转到登录页面
        if (session.getAttribute("user")==null)
            response.sendRedirect("login.jsp");
    %>
    <div class="well">
        如果您没有合法的身份验证,则无法看到此页,能够看到此页,说明您已经通过了合法的身份验证!
    </div>
    <div class="well">
        如果您选择了记住密码,下次登录时就无须输入账号与密码,返回<a href="login.jsp">重新登录</a>,验证一下。
    </div>
    <div class="well ">
        <%
            /*访问历史人数写入*/
            if (application.getAttribute("count")==null)
                application.setAttribute("count",1);
            else if (session.isNew())
                application.setAttribute("count",(Integer)application.getAttribute("count")+1);
            out.print("您是当前第【"+(Integer)application.getAttribute("count")+"】位访问网站的用户!");
        %>
    </div>
    </body>
</html>
```

运行结果

用户登录页面时,首先运行 login.jsp 页面,效果如图 5-9 所示。

图 5-9 用户登录页面

单击图 5-9 中的"登录"按钮后,通过 checkLogin.jsp 页面进行登录验证,验证通过后,将跳转至 index.jsp 页面,效果如图 5-10 所示。

用户身份验证案例展示了 request、session、response、out 和 application 等 5 个内置对象的综合应用,可以将这些技术直接运用到实际的项目中,也可以在此基础上进行功能的扩充和优化。其他的内置对象由于使用频率不高,不再做案例展示。

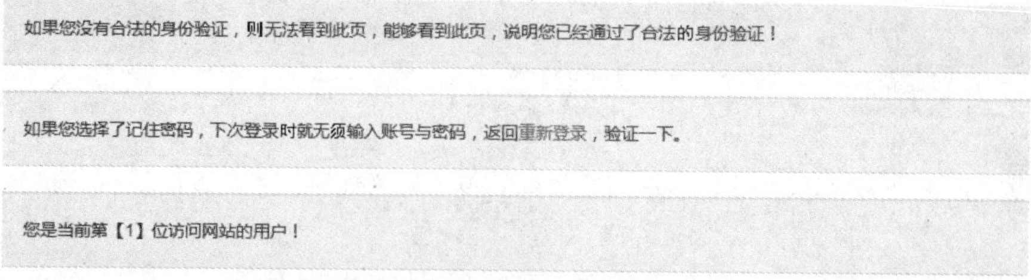

如果您没有合法的身份验证，则无法看到此页，能够看到此页，说明您已经通过了合法的身份验证！

如果您选择了记住密码，下次登录时就无须输入账号与密码，返回重新登录，验证一下。

您是当前第【1】位访问网站的用户！

图 5-10 index.jsp 主页面

习题

利用内置对象技术实现对学生信息的维护，项目具体要求如下：
1．学生属性：学号（唯一性）、姓名、性别及班级；
2．学生信息维护功能包括学生信息的新增、修改、删除及查找；
3．采集到的学生信息数据，全部保存在 application 内置对象中；
4．学生信息维护前，需要进行身份验证，模拟登录用户名 admin，密码 admin；如果验证不成功，则需要重新验证；
5．采集到的学生信息实时显示在页面上，页面风格不限；
6．学生信息的新增、修改、删除及查找页面风格不限。

第 6 章

EL 与 JSTL 标签

EL 与 JSTL 标签是一种将 Java 运行结果输出到页面上的技术。JSP 中 EL 是表达式语言（Expression Language）的缩写，利用它可以通过简单的编写方式将 JavaBean 和内置对象中的数据呈现在 JSP 页面上。JSTL（JavaServer Pages Standard Tag Library，JSP 标准标签库）是一个不断完善的开源 JSP 标签库，由 apache 的 jakarta 小组来维护。JSTL 只能运行在支持 JSP 1.2 和 Servlet 2.3 规范的容器上，在 JSP 2.0 中也作为标准支持。本章将详细讲解 EL 的语法和 JSTL 的核心库标签语法，通过项目案例的方式阐述 EL 和 JSTL 标签库在项目中的应用。

本章任务

（1）EL 表达式编写规则；
（2）JSTL 核心库标签的使用规则；
（3）EL 与 JSTL 项目应用。

重点内容

（1）掌握 EL 表达式的编写规范；
（2）掌握 JSTL 核心库标签的使用；
（3）掌握 EL 与 JSTL 项目的应用。

难点内容

（1）JSTL 核心库标签的使用；
（2）EL 与 JSTL 项目应用。

6.1 EL 与 JSTL 概述

本节详细阐述了 EL 与 JSTL 的相关概念及相关语法，包括 EL 表达式语言编写规则和

JSTL 核心库标签的使用规范。

6.1.1　EL 概述

EL（Expression Language）表达式语言，它简化了 Java 服务器端输出数据在页面上的编写方式，通过其规定的简单语法规范就可以将部分内置对象中的值或其自定义表达式的运算结果输出在页面指定的标签上。

EL 自定义表达式中，可以运用算术表达式、关系运算表达式和逻辑表达式，在其表达式内，可以使用整型数、浮点数、字符串、常量 true、false 及 null 等类型的值。

EL 表达式的语法格式：${表达式}

语法中：必须以"$"开头，再使用大括号"{}"，表达式必须写在大括号内，表达式由各种操作符和操作数组成。在 EL 表达式中，通用操作符是"."。这个操作符允许通过内嵌 JSP 对象访问其 JavaBean 属性。

EL 表达式可以获取部分内置对象中的值，具体包括：

1. request

语法格式：${requestScope.属性名}

说明：若使用${requestScope.属性名}表达式，需要内置 request 对象通过 request.setAttribute(属性名,属性值)设置属性值；属性名要与 requestScope.属性名一致。

2. session

语法格式：${sessionScope.属性名}

说明：若使用${sessionScope.属性名}表达式，需要内置 session 对象通过 session.setAttribute(属性名,属性值)设置属性值；属性名要与 sessionScope.属性名一致。

3. application

语法格式：${applicationScope.属性名}

说明：若使用${applicationScope.属性名}表达式，需要内置 application 对象通过 application.setAttribute(属性名,属性值)设置属性值；属性名要与 applicationScope.属性名一致。

4. pageContext

语法格式：${pageScope.属性名}

说明：若使用${pageScope.属性名}表达式，需要内置 pageContext 对象通过 pageContext.setAttribute(属性名,属性值)设置属性值；属性名要与 pageScope.属性名一致。

EL 表达式除可以获取部分内置对象中的值外，还可以获取页面传递过来的参数和 cookie 中的数据。

1. 获取页面传递参数值

语法格式：${param.参数名}

说明：${param.参数名}中的参数名就是表单中输入标签 name 属性的值。

2. 获取 cookie 值

语法格式：${cookie.属性名}

说明：${cookie.属性名}中的属性名要与写入时的属性名一致，写入 cookie 数据时，可以通过 response.addCookie(newCookie(属性名,属性值))格式写入。

EL 表达式可以自定义编写表达式，支持算术表达式、关系运算表达式和逻辑运算表达式，具体的运算符及说明参照表 6-1。

表 6-1　EL 表达式运算符及说明

序　号	操 作 符	说　　明
1	.	点运算符，通过它可以访问 JavaBean 对象的属性值
2	[ ]	数组运算，通过它可以访问数组中的指定元素
3	( )	括号运算符，可以改变运算符的优先级
4	+	加法运算符
5	-	减法运算符
6	*	乘法运算符
7	/或者 div	除法运算符，可以用"/"或者"div"来表示
8	%或者 mod	求余运算符，可以用"%"或者"mod"来表示
9	==或者 eq	关系运算符，等于
10	!=或者 ne	关系运算符，不等于
11	<或者 lt	关系运算符，小于
12	>或者 gt	关系运算符，大于
13	<=或者 le	关系运算符，小于或等于
14	>=或者 ge	关系运算符，大于或等于
15	&&或者 and	逻辑运算符，与运算符
16	||或者 or	逻辑运算符，或运算符
17	empty	判断某个对象是否为空，为空返回 true，否则返回 false

6.1.2　JSTL 概述

JSTL（JavaServer Pages Standard Tag Library，JSP 标准标签库）是一个不断完善的开源 JSP 标签库，通过 JSTL 标签库与 EL 的结合将 Java 服务器端的数据呈现在页面上。

JSTL 标签库可分为核心标签库（core）、格式化标签库（format）、SQL 标签库和 XML 标签库，如图 6-1 所示。

- 核心标签库提供了定制操作，通过限制作用域的使用范围来管理数据，以及执行页面内容的迭代和条件来操作数据。它还提供了用来生成和操作 URL 的标记；
- 格式化标签库定义了用来格式化数据（尤其是数字和日期）的操作。它还支持使用本地化资源库进行 JSP 页面的国际化；
- XML 标签库包含一些标记，这些标记用来操作通过 XML 表示的数据；
- SQL 标签库定义了用来查询关系数据库的操作。

图 6-1　JSTL 标签库分类

JSTL 标签库需要在 JSP 页面中使用，在使用前，须进行必要的配置，JSTL 需要

standard.jar 和 jstl.jar 两个 jar 包。jar 包可以通过 appache 官网下载,再将这两个包加载到所需的项目中。

JSTL 的核心标签库是最常用的标签,使用时需要在 JSP 页面头部使用编译指令<taglib>引入标签,语法格式:

```
<%@ taglib prefix="c" uri="http://java.sun.com/jsp/jstl/core" %>
```

在 JSTL 核心标签库中,包含的具体标签见表 6-2。

表 6-2　JSTL 的核心标签库说明

序　号	标 签 名	说　　明
1	<c:out>	将指定的值向页面输出
2	<c:set>	设置变量并初始化
3	<c:remove>	移除指定变量,移除的变量中的值也会消失
4	<c:catch>	处理错误异常信息,并将这些异常错误信息保存起来
5	<c:if>	判断语句,与 Java 中的 if 功能类似
6	<c:choose>	多分支选择结构语句,类似于 Java 中的 switch 语句
7	<c:when>	多分支选择结构语句的一个分支
8	<c:otherwise>	多分支选择结构中除 when 之外的情况
9	<c:import>	引入其他网页资源,类似于编译指令<jsp:include>,并保存
10	<c:forEach>	循环遍历功能,类似于 Java 中的 for 和 while 功能
11	<c:forToken>	循环结构,通过指定分隔符将字符串分割成一个数组后进行迭代
12	<c:param>	设置<c:url>地址中所需的地址参数
13	<c:redirect>	页面重定向到一个新的 URL 地址
14	<c:url>	定义一个页面跳转地址

1. <c:out>标签

<c:out>是页面输出标签,其功能是将指定的数值输出到 JSP 页面上,类似于 JSP 内置对象 out。

语法格式:

```
<c:out
    value="输出的值"
    default="输出默认值"
    escapeXML="true|false">
</c:out>
```

<c:out>标签中各种属性说明见表 6-3。

表 6-3　<c:out>标签属性说明

属　　性	描　　述	是否必要	默 认 值
value	输出的内容,可以是直接字符串,也可以是 EL 表达式的值	是	无
default	输出的默认值	否	
escapeXML	是否忽略输出内容 value 中的 XML 特殊字符,true 表示忽略,false 表示不忽略。例如:如果要输出原版的 HTML 标签而不被浏览器解析,就需要设置该值为 true 或者不设置	否	true

2. <c:set>标签

JSTL 中的<c:set>标签用于定义变量,并且进行初始化。

语法格式:
```
<c:set
    var="变量名"
    value="初始化值"
    target="JavaBean对象引用对象名称"
    property="引用JavaBean对象属性名"
    scope="应用范围" >
</c:set>
```
<c:set>标签中的各种属性说明见表 6-4。

表 6-4 <c:set>标签属性说明

属性	描述	是否必要	默认值
var	定义变量,变量名遵循标识符命名规则即可	否	无
value	定义变量的初始化值	否	无
target	引用的 JavaBean 对象名称,需要使用"${对象名称}"格式引用	否	无
property	用于修改引用 JavaBean 对象的属性值,与 target 联合使用	否	无
scope	变量的作用域	否	当前页面

属性 target 和 property 常在一起使用。例如:有 User 的 JavaBean 类,含有账号 number 属性,通过定义<c:set>标签将账号 number 属性值设置为"admin"。

第一步:通过动作指令 jsp:useBean 创建 User 的 Bean 对象 u:
```
<jsp:useBean id="u" class="User"></jsp:useBean>
```
第二步:利用 JSTL 中<c:set>标签设置对象 u 的 number 属性值:
```
<c:set target="${u}" property="number" value="admin" ></c:set>
```

3. **<c:remove>标签**

<c:remove>标签用来移除<c:set>标签定义的变量。
语法格式:
```
<c:remove var="移除的变量名称" scope="作用域"></c:remove>
```
<c:remove>属性说明参照表 6-5。

表 6-5 <c:remove>属性说明

属性	描述	是否必要	默认值
var	要移除<c:set>标签中已经定义的变量名称	是	无
scope	作用域范围,可以是 page、scope、application 和 session	否	所有作用域

4. **<c:catch>标签**

<c:catch>标签用于处理产生错误异常状况,并进行存储,功能类似于 Java 中的 try…catch。
语法格式:
```
<c:catch var="存储异常信息变量名称"></c:catch>
```

5. **<c:if>标签**

<c:if>标签判断语句只有其表达式为 true 时才会执行,功能与 Java 中的 if 语句类似,但没有 else 分支。

语法格式：
```
<c:if
 test="表达式1"
 var="变量名"
 scope="作用域">
</c:if>
```

<c:if>标签中各属性说明请参照表 6-6。

表 6-6 <c:if>属性说明

属 性	描 述	是否必要	默 认 值
test	判断条件或条件表达式，结果是布尔值	是	无
var	存储 test 属性结果的变量名称	否	无
scope	var 的作用域范围，可以是 page、scope、application 和 session	否	page

6. 多分支选择结构标签

多分支选择结构标签由<c:choose>、<c:when>、<c:otherwise>三个标签组成，其功能与 Java 中的 switch 语句类似。

语法格式定义：
```
<c:choose>
    <c:when test="表达式或布尔值">
        ...
    </c:when>
    <c:when test="表达式或布尔值">
        ...
    </c:when>
    ...
    ...
    <c:otherwise>
        ...
    </c:otherwise>
</c:choose>
```

<c:choose>标签没有属性，<c:when>标签有一个 test 属性，表示条件成立时要处理的动作，<c:otherwise>标签表示所有的<c:when>都不成立时执行的动作语句。

7. <c:import>标签

<c:import>标签类似于<jsp:include>动作指令功能，可以引入相应的页面资源到当前页面中。

语法格式：
```
<c:import
    url="资源地址"
    var="变量名称"
    scope="var的作用范围"
    varReader="提供给java.io.Reader对象的变量名称"
    context="资源名称"
    charEncoding="编码格式">
</c:import>
```

<c:import>标签中各属性的使用规则见表 6-7。

表 6-7 <c:import>属性使用规则

属　性	描　述	是否必要	默 认 值
url	导入资源的路径，可以是相对路径或绝对路径	是	无
context	当使用相对路径访问外部 context 资源时，使用 context 设置资源名称	否	当前应用程序
charEncoding	所引入资源字符编码集	否	ISO-8859-1
var	存储所引入 URL 的变量	否	无
scope	var 属性的作用域	否	page
varReader	用于提供 java.io.Reader 对象的变量	否	无

8. <c:foreach>标签

<c:foreach>标签类似于 Java 中的 while 或 for 循环功能，对指定的集合进行迭代遍历。
语法格式：

```
<c:foreach
   items="集合对象"
   begin="循环开始下标"
   end="结束下标"
   step="循环步长"
   var="变量名称"
   varStatus="变量状态名称">
</c:foreach>
```

<c:foreach>标签的各种属性说明见表 6-8。

表 6-8 <c:foreach>标签属性说明

属　性	描　述	是否必要	默 认 值
items	遍历的集合对象	否	无
begin	开始的元素（0=第一个元素）	否	0
end	最后一个元素	否	最后一个元素
step	循环步长	否	1
var	变量名称，存放当前遍历的对象	否	无
varStatus	代表循环状态的变量名称	否	无

其中，varStatus 属性命名的变量并不存储当前索引值或当前元素，而是赋予 javax.Servlet.jsp.jstl.core.LoopTagStatus 类的实例。该类包含了一系列的特性，它们描述了迭代的当前状态，其属性的含义见表 6-9。

表 6-9 varStatus 的属性含义

属　性	含　义
current	当前正在迭代（集合中的）的项
index	当前正在迭代的索引
count	当前已经迭代的总数
first	当前迭代是否为第一次迭代，该属性为 boolean 型
last	当前迭代是否为最后一次迭代，该属性为 boolean 型
begin	begin 属性的值
end	end 属性的值
step	step 属性的值

9. <c:forTokens>标签

<c:forTokens>标签通过指定的分隔符号将目标字符串分割成一个数组，然后进行迭代。
语法格式：

```
<c:forTokens
    items="集合对象"
    begin="循环开始下标"
    end="结束下标"
    step="循环步长"
    var="变量名称"
    delims="分割字符"
    varStatus="变量状态名称">
</c:forTokens>
```

标签<c:forTokens>的语法与<c:forEach>基本一致，区别在于<c:forTokens>标签增加了属性 delims，表示用于字符串分割的字符。

10. <c:url>与<c:param>标签

<c:url>标签用于定义 URL 地址，并将定义的结果存储在指定变量中，在定义 URL 地址时，可以带参数，参数采用<c:param>标签进行定义。
语法格式：

```
<c:url
    var="变量名"
    scope="作用域"
    value="URL地址"
    context="应用程序名称">
    <c:param name="参数名" value="参数值"/>
    …
</c:url>
```

说明，在<c:url>标签中，参数标签<c:param>不是必要项，但是<c:param>标签必须放在<c:url>中使用。<c:url>标签的相关属性说明见表 6-10。

表 6-10 <c:url>标签属性说明

属　　性	描　　述	是否必要	默　认　值
value	URL 地址	是	无
context	应用程序的名称	否	当前应用程序
var	存储 URL 的变量名	否	当前页面
scope	var 属性的作用域	否	Page

11. <c:redirect>标签

<c:redirect>标签通过重写 URL 来将浏览器跳转到一个新的 URL，并且支持<c:param>标签。
语法格式：

```
<c:redirect url="跳转目标地址" context="当前应用程序名称"/>
```

6.2 剪刀石头布游戏

通过剪刀石头布的经典游戏掌握 EL 与 JSTL 中<c:choose>标签的应用。

案例描述

剪刀石头布是一款经典的游戏，本案例要求利用 EL 与 JSTL 中的选择标签将这款经典游戏实现人机交互；具体要求如下：
- 提供一个用户选择"剪刀、石头或布"的页面，并提交到服务器；
- 服务器随机产生"剪刀、石头或布"；
- 服务器将用户选择的"剪刀石头布"游戏的结果与服务器随机产生的剪刀石头布结果进行比对；
- 新页面显示人机比赛结果；
- 要求实现上述功能时，尽可能使用 EL 与 JSTL 标签进行页面呈现。

案例分析

根据案例描述，开发一款经典"剪刀石头布"游戏主要考察 EL 与 JSTL 中的多分支判断标签的使用，游戏分为页面呈现和程序实现两部分，页面呈现设计如图 6-2 所示。

图 6-2 "剪刀石头布"游戏页面设计

图 6-2 中，将游戏页面设计分为了两个页面："玩家页面"提供玩家选择"剪刀石头布"中的任何一个，选择完毕后，单击"开始比赛"按钮，系统将玩家选择的结果提交到"游戏结果页面"，此页面实现的不仅仅是将比赛及结果呈现出来，还需要计算机随机产生"剪刀石头布"中的任意一个，再与人机选择的结果比对，并将比对结果呈现在页面上。服务器端随机产生"剪刀石头布"的结果与玩家结果比较的业务逻辑设计如图 6-3 所示。

图 6-3 中，首先通过玩家页面获取用户选择"剪刀、石头、布"中的一种结果，提交到"游戏结果页面"，"游戏结果页面"先要处理计算机端的程序，再选择由计算机随机产生"剪刀、石头、布"中的随机结果，并与玩家提供的结果进行比对，图 6-3 中清晰地展示了 9 种人机比对结果，最后将比对的结果呈现在页面上。

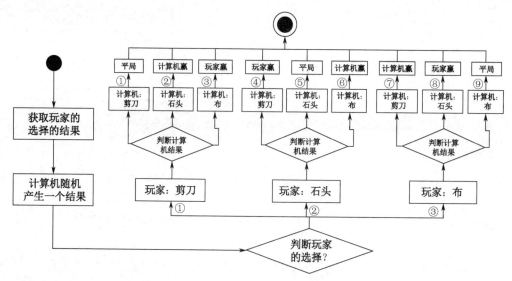

图 6-3 剪刀石头布业务逻辑设计

案例实现

根据案例分析，页面部分由"玩家页面"和"游戏结果页面"组成。

1. 玩家页面 game.jsp

玩家页面 game.jsp 提供了 3 个单选按钮，分别代表剪刀、石头、布，以及一个"开始比赛"按钮，通过表单管理这些输入标签，具体页面实现代码如下：

```jsp
<%--
  Created by IntelliJ IDEA.
  User: wph-pc
  Date: 2018/12/25
  Time: 20:44
  To change this template use File | Settings | File Templates.
--%>
<%@ page contentType="text/html;charset=UTF-8" language="java" %>
<html>
<head>
    <title>剪刀石头布游戏</title>
</head>
<body>
    <h2>剪刀石头布"人机"游戏</h2>
    <hr>
    <form method="post" action="doGame.jsp">
        <h3>请选择</h3>
        <div>
            <input type="radio" name="rbGame" value="scissor">剪刀  
            <input type="radio" name="rbGame" value="stone">石头  
            <input type="radio" name="rbGame" value="cloth">布  
        </div>
        <input type="submit" value="开始比赛">
    </form>
</body>
```

```
</html>
```

以上代码中,当单击"开始比赛"按钮后,页面将跳转到 doGame.jsp,此页面实现游戏结果显示。

2. 游戏结果页面 doGame.jsp

doGame.jsp 页面不仅显示游戏比赛结果,还需要实现计算机端的"剪刀石头布"的随机产生和比对,最后将比对结果呈现在页面上。随机产生"剪刀石头布"采用了随机数策略,0 表示剪刀,1 表示石头,2 表示布。核心代码如下:

```java
//随机数,0表示剪刀,1表示石头,2表示布
int seed=0;
Random rand=new Random();
//产生随机数后,取3的余数
seed=rand.nextInt(100)%3;
```

在人机比对部分,采用了 JSTL 标签库中的<c:choose>多分支选择标签,具体页面实现代码如下:

```jsp
<%@ page import="java.util.Random" %><%--
  Created by IntelliJ IDEA.
  User: wph-pc
  Date: 2018/12/25
  Time: 20:50
  To change this template use File | Settings | File Templates.
--%>
<%@ page contentType="text/html;charset=UTF-8" language="java" isELIgnored="false" %>
<%@ taglib prefix="c" uri="http://java.sun.com/jsp/jstl/core" %>
<html>
<head>
    <title>处理剪刀石头布游戏</title>
</head>
<body>
<%
    //随机数,0表示剪刀,1表示石头,2表示布
    int seed=0;
    Random rand=new Random();
    //产生随机数后,取3的余数
    seed=rand.nextInt(100)%3;
%>
<c:set var="s" value="<%=seed%>"></c:set>
<c:choose>
    <c:when test="${param.rbGame=='stone'}">
        <c:choose>
            <c:when test="${s==0}">
                <c:out value="计算机出:剪刀;<br>您出:石头;<br>您【赢】" escapeXml="false"></c:out>
            </c:when>
            <c:when test="${s==1}">
                <c:out value="计算机出:石头;<br>您出:石头;<br>平局" escapeXml="false"></c:out>
            </c:when>
```

```
            <c:when test="${s==2}">
                <c:out value="计算机出：布；<br>您出：石头；<br>计算机【赢】" escapeXml="false"></c:out>
            </c:when>
            <c:otherwise>
                <c:out value="计算机数据异常！"></c:out>
            </c:otherwise>
        </c:choose>
    </c:when>
    <c:when test="${param.rbGame=='scissor'}">
        <c:choose>
            <c:when test="${s==0}">
                <c:out value="计算机出：剪刀；<br>您出：剪刀；<br>平局" escapeXml="false"></c:out>
            </c:when>
            <c:when test="${s==1}">
                <c:out value="计算机出：石头；<br>您出：剪刀；<br>计算机【赢】" escapeXml="false"></c:out>
            </c:when>
            <c:when test="${s==2}">
                <c:out value="计算机出：布；<br>您出：剪刀；<br>您【赢】" escapeXml="false"></c:out>
            </c:when>
            <c:otherwise>
                <c:out value="计算机数据异常！"></c:out>
            </c:otherwise>
        </c:choose>
    </c:when>
    <c:when test="${param.rbGame=='cloth'}">
        <c:choose>
            <c:when test="${s==0}">
                <c:out value="计算机出：剪刀；<br>您出：布；<br>计算机【赢】" escapeXml="false"></c:out>
            </c:when>
            <c:when test="${s==1}">
                <c:out value="计算机出：石头；<br>您出：布；<br>您【赢】" escapeXml="false"></c:out>
            </c:when>
            <c:when test="${s==2}">
                <c:out value="计算机出：布；<br>您出：布；<br>平局" escapeXml="false"></c:out>
            </c:when>
            <c:otherwise>
                <c:out value="计算机数据异常！"></c:out>
            </c:otherwise>
        </c:choose>
    </c:when>
    <c:otherwise>
        <c:out value="您没有出任何动作！"></c:out>
    </c:otherwise>
</c:choose>
```

```
            </body>
         </html>
```

页面 doGame.jsp 中，使用了 EL 表达式和 JSTL 中的<c:set>、<c:out>、<c:choose>、<c:when>、<c:otherwise>等标签，该案例详细讲解了这些标签的综合应用。

运行结果

运行案例中的 game.jsp，结果如图 6-4 所示。图 6-4 中，提供了"剪刀、石头、布"3 个单选按钮，玩家需要从"剪刀、石头或布"中选择任意一项，单击"开始比赛"按钮，将跳转到 doGame.jsp 页面，如图 6-5 所示。

图 6-4 "剪刀石头布"玩家页面

图 6-5 "剪刀石头布"游戏结果显示

6.3 发牌游戏

纸牌可以有很多种玩法，但不管怎么玩，需要先洗牌，本案例通过对一副纸牌进行洗牌，分四组进行发牌，这些过程都由计算机完成。通过本案例，重点学习 JSTL 标签库中<c:forEach>标签的应用。

案例描述

利用 JSP 技术，产生一副 54 张纸牌，再随机洗牌打乱，最后模拟四人进行发牌。具体要求如下：

- 由 Java 程序产生一副纸牌，1~10 用数字字符表示，J、Q、K 用字母字符表示，每张纸牌需要标明纸牌的数值和花色；还需要两张"鬼牌"（大王和小王）；
- 由 Java 程序将产生的纸牌随机洗牌 10000 次；
- 发牌，分成四组，从第一组开始依次发放；
- 发牌时按照分组显示在页面上，尽量提高用户的 UI 体验；
- 尽可能采用 JSTL 标签技术呈现纸牌。

案例分析

根据案例描述，须解决以下问题：

- 如何表示纸牌的数值及花色；
- 纸牌的产生及洗牌；

- 纸牌的发放；
- 纸牌的呈现。

针对上述问题，设计了如图 6-6 所示的解决思路：

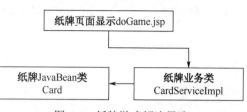

图 6-6 纸牌游戏解决思路

图 6-6 中，根据提出的四个待解决的问题设计了相应的解决办法，纸牌的数值及花色表示由纸牌 JavaBean 类 Card 实现；纸牌的产生及洗牌由纸牌业务类 CardServiceImpl 实现；而纸牌的发放及呈现由 doGame.jsp 页面实现。

1. 纸牌 Card 类

纸牌 Card 类是一个 JavaBean 类，纸牌具体属性的设计如表 6-11 所示。

表 6-11 Card 类

类 名 称	Card	中文名称	纸牌	类 型	实体类
类 描 述	提供纸牌基本信息，属于一个 JavaBean 类				
属 性					
序 号	属 性 名	权限与修饰词	类 型	说 明	
1	name	private	String	纸牌的数值	
2	color	private	String	纸牌的花色	
方 法					
序 号	方 法 名	参 数	返回类型	功能说明	
1	getName	无	String	获取纸牌数值名称	
2	setName	String name：纸牌名称	void	设置纸牌名称	
3	getColor	无	String	获取纸牌花色	
4	setColor	String color：纸牌花色	void	设置纸牌花色	

2. 纸牌业务类设计

纸牌业务类（CardServiceImpl）主要是实现纸牌的产生和洗牌功能，具体的设计见表 6-12。

表 6-12 CardServiceImpl 类

类 名 称	CardServiceImpl	中文名称	纸牌业务类	类 型	实体类
类 描 述	提供纸牌功能操作，包括纸牌的产生和洗牌功能				
方 法					
序 号	方 法 名	参 数	返回类型	功能说明	
1	getPokerCardInfo	int data：纸牌数值转字符。data 整数	String	私有方法，将整数参数转换成字符串，特别是 J、Q、K 的处理	
2	createCards	无	List<Card>	公共方法，实现 54 张纸牌的产生，并返回一副完整纸牌	
3	mixCards	List<Card> cards	void	公共方法，洗牌功能，实现 10000 次洗牌	

纸牌业务类 CardServiceImpl 中的洗牌（mixCards）功能复杂点，为了更好地实现该功

能，图 6-7 展示了该功能的程序设计过程：

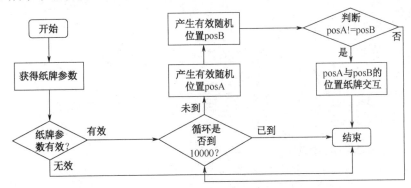

图 6-7 洗牌功能实现设计

图 6-7 中，先获取一副纸牌参数，判断该纸牌是否存在，如果不存在，就结束；如果存在，则进入 10000 次的洗牌循环。每次循环时，随机产生 posA 和 posB 位置，该位置为纸牌集合的索引位置，如果 posA 不等于 posB 的位置，则对这两个位置的纸牌进行交换，如此循环，直到 10000 次循环结束。

案例实现

根据案例分析，该案例实现由纸牌类 Card、纸牌业务类 CardServiceImpl 类、发牌及洗牌 cardGame.jsp 页面组成。

1. Card 类

根据案例分析中的 Card 类的设计，具体实现如下：

```java
package chapter6;

/**
 * 纸牌实体类
 * Created by wph-pc on 2018/12/25.
 */
public class Card {
//名称
    private String name="";
    //花色
    private String color="";

    public String getName() {
        return name;
    }
    public void setName(String name) {
        this.name = name;
    }
    public String getColor() {
        return color;
    }
    public void setColor(String color) {
        this.color = color;
```

            }
        }

2. 纸牌业务类 CardServiceImpl

根据纸牌业务类的设计，纸牌业务类的代码实现如下：

```java
package chapter6;

import java.util.ArrayList;
import java.util.Iterator;
import java.util.List;
import java.util.Random;

/**
 * 纸牌功能类，实现创建纸牌、洗牌
 * Created by wph-pc on 2018/12/25.
 */
public class CardServiceImpl {
    /*
     *将数字data转成符合要求的J、Q或K
     * 数字11代表J，12代表Q，13代表K，其他都返回数字本身
     * */
    private String getPokerCardInfo(int data){
        switch (data){
            case 11:
                return "J";
            case 12:
                return "Q";
            case 13:
                return "K";
            default:
                return String.valueOf(data);
        }
    }
    /*创建一副纸牌*/
    public List<Card> createCards(){
        //定义存放纸牌临时变量temp
        List<Card> temp=new ArrayList<>();
        /*创建52张不同颜色的数字牌存入temp中*/
        for(int i=1;i<=13;i++){
            //红色牌对象
            Card cardHeart=new Card();
            cardHeart.setColor("红桃");
            cardHeart.setName(getPokerCardInfo(i));
            temp.add(cardHeart);

            //黑色牌对象
            Card cardSpade=new Card();
            cardSpade.setColor("黑桃");
            cardSpade.setName(getPokerCardInfo(i));
            temp.add(cardSpade);
```

```
        //方块牌对象
        Card cardDiamond=new Card();
        cardDiamond.setColor("方块");
        cardDiamond.setName(getPokerCardInfo(i));
        temp.add(cardDiamond);

        //梅花牌对象
        Card cardClubs=new Card();
        cardClubs.setColor("梅花");
        cardClubs.setName(getPokerCardInfo(i));
        temp.add(cardClubs);
    }
    /*增加一张大王*/
    Card cardKing=new Card();
    cardKing.setColor("无");
    cardKing.setName("大王");
    temp.add(cardKing);
     /*增加一张小王*/
    Card cardQueen=new Card();
    cardQueen.setColor("无");
    cardQueen.setName("小王");
    temp.add(cardQueen);
    //返回创建54张牌
    return temp;
}

/*将一幅纸牌进行随机洗牌*/
public void mixCards(List<Card> cards){
    //判断纸牌有效性
    if (cards==null || cards.size()==0) return;
    /*随机洗牌10000次*/
    for(int i=0;i<10000;i++){
        Random rand=new Random();
        //产生第一个随机数位置，0~53之间的整数
        int posA=rand.nextInt(1000)%54;
        //产生第二个随机数位置，0~53之间的整数
        int posB=rand.nextInt(1000)%54;
        if(posA!=posB){
            Card temp=cards.get(posA);
            cards.set(posA,cards.get(posB));
            cards.set(posB,temp);
        }
    }
}
```

3. 纸牌发放及显示页面 cardGame.jsp

根据案例分析中的设计，实现纸牌发放及呈现由 cardGame.jsp 页面完成，洗完后的纸牌数据提取及发放主要由 JSTL 中的<c:forEach>、<c:if>和<c:choose>标签实现。在纸牌呈现时，为了提升用户体验，页面使用了4种不同花色的图片和2张表示大小王的图片，该

页面的具体实现如下：

```jsp
<%@ page import="chapter6.CardServiceImpl" %>
<%@ page import="chapter6.Card" %>
<%@ page import="java.util.List" %><%--
  Created by IntelliJ IDEA.
  User: wph-pc
  Date: 2018/12/26
  Time: 19:35
  To change this template use File | Settings | File Templates.
--%>
<%@ page contentType="text/html;charset=UTF-8" language="Java" isELIgnored="false" %>
<%@ taglib prefix="c" uri="http://java.sun.com/jsp/jstl/core" %>
<%@ include file="../header/init_bootstrap.jsp"%>
<html>
<head>
<title>发牌程序，利用JSTL实现发牌</title>
    <style>
        .card{
            width:40px;
            height:50px;
            border: solid 1px grey;
            border-radius: 5px;
        }
    </style>
</head>
<body>
    <!--创建纸牌，并洗牌-->
    <%
        //创建发牌业务对象
        CardServiceImpl cardBll=new CardServiceImpl();
        //创建纸牌
        List<Card> cards=cardBll.createCards();
        //洗牌
        cardBll.mixCards(cards);
    %>
    <c:set value="<%=cards%>" var="cards"></c:set>
    <h2>发牌程序</h2>
    <hr>
    <div>服务器端随机产生一副纸牌，分四组发放</div>
    <table class="table">
        <thead>
        <tr>
            <th>第一组</th>
            <th>第二组</th>
            <th>第三组</th>
            <th>第四组</th>
        </tr>
        </thead>
        <tbody>
        <tr><td colspan="4">纸牌从第一组到第四组循环发放</td>
```

```jsp
                <c:forEach items="${cards}" begin="0" varStatus="status">
            <c:if test="${status.index%4==0}" var="flag">
                </tr><tr>
                    <c:if test="${status.index<status.count}">
                        <c:choose>
                            <c:when test="${cards[status.index].color=='红桃'}">
                                <td><img src="../images/heartCard.png" class="card">${cards[status.index].name}</td>
                            </c:when>
                            <c:when test="${cards[status.index].color=='方块'}">
                                <td><img src="../images/diamondCard.png" class="card">${cards[status.index].name}</td>
                            </c:when>
                            <c:when test="${cards[status.index].color=='黑桃'}">
                                <td><img src="../images/spadeCard.png" class="card">${cards[status.index].name}</td>
                            </c:when>
                            <c:when test="${cards[status.index].color=='梅花'}">
                                <td><img src="../images/clubsCard.png" class="card">${cards[status.index].name}</td>
                            </c:when>
                            <c:when test="${cards[status.index].name=='大王'}">
                                <td><img src="../images/bigKing.png" class="card"></td>
                            </c:when>
                            <c:when test="${cards[status.index].name=='小王'}">
                                <td><img src="../images/smallKing.png" class="card"></td>
                            </c:when>
                            <c:otherwise></c:otherwise>
                        </c:choose>
                    </c:if>
            </c:if>
            <c:if test="${flag==false}">
                <c:if test="${status.index<status.count}">
                    <c:choose>
                        <c:when test="${cards[status.index].color=='红桃'}">
                            <td><img src="../images/heartCard.png" class="card">${cards[status.index].name}</td>
                        </c:when>
                        <c:when test="${cards[status.index].color=='方块'}">
                            <td><img src="../images/diamondCard.png" class=
```

```
"card">${cards[status.index].name}</td>
                                    </c:when>
                                    <c:when test="${cards[status.index].color=='黑桃'}">
                                    <td><img src="../images/spadeCard.png" class="card">${cards[status.index].name}</td>
                                    </c:when>
                                    <c:when test="${cards[status.index].color=='梅花'}">
                                    <td><img src="../images/clubsCard.png" class="card">${cards[status.index].name}</td>
                                    </c:when>
                                    <c:when test="${cards[status.index].name=='大王'}">
                                    <td><img src="../images/bigKing.png" class="card"></td>
                                    </c:when>
                                    <c:when test="${cards[status.index].name=='小王'}">
                                    <td><img src="../images/smallKing.png" class="card"></td>
                                    </c:when>
                                    <c:otherwise></c:otherwise>
                                </c:choose>
                            </c:if>
                        </c:if>
                    </c:forEach>
                </tbody>
            </table>

    </body>
</html>
```

运行结果

启动运行案例中的 cardGame.jsp 页面，运行结果如图 6-8 所示。

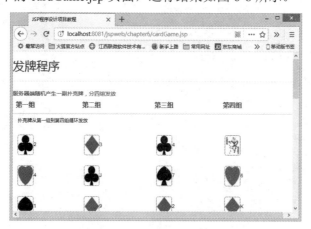

图 6-8 纸牌发放及显示

图 6-8 中，页面 cardGame.jsp 将纸牌业务类产生的纸牌随机洗牌后，通过 JSTL 中的 <c:forEach>、<c:if>和<c:choose>等标签分四组顺序循环发放，每次从第一组开始，到第四组结束，以此循环，直到纸牌发放完毕。纸牌在呈现时，使用一些图片代替不同花色及两种"鬼牌"。

习题

在 JSP 页面上显示商品信息，每个商品信息包括：商品编号、商品名称、商品图片、单价、商品介绍等基本信息，具体要求如下：

1．在服务器端产生 10 个商品信息，读者自行定义即可，存放在 request 内置对象中；

2．新建一个 JSP 页面用来分别采用 table 表格及列表两种方式显示服务器端的 10 个商品信息；

3．要求在读取服务器端的数据时，尽可能多采用 JSTL 与 EL 技术。

第 7 章

JavaBean 技术

JavaBean 技术在项目开发中的应用非常广泛，它是一种特殊的 Java 类，其他 Java 类可以通过自省机制（反射机制）发现和操作该类的属性，包括对这些属性进行赋值或获取属性值。

本章任务

（1）掌握 JavaBean 类的编写规范；
（2）利用 JavaBean 类进行赋值及取值应用。

重点内容

（1）JavaBean 类的编写规范；
（2）JavaBean 的项目应用。

难点内容

JavaBean 的项目应用。

7.1　JavaBean 概述

JavaBean 是一种特殊的 Java 类，它必须是具体的和公共的，并且具有无参数的构造器。具体要求：

- 构造方法没有参数
- 属性必须私有化
- 私有化属性访问可通过含有 public 修饰符的 setXXX 或 getXXX 方法。setXXX 方法用于设置属性值，getXXX 方法用于读取属性值。其中的"XXX"表示属性名称，首写字母大写，例如：属性名称为"name"，设置 name 属性值的方法名为

setName，获取属性值的方法名为 getName。

7.1.1 JavaBean 组成

一个 JavaBean 类一般由属性、方法及事件组成。

1. 属性

JavaBean 属性值的读写可以通过调用适当的 bean 方法进行。比如，JavaBean 类有一个 name 属性，此属性值可通过调用 String getName()方法读取，而写入属性值则须调用 void setName(String name)方法。

每个 JavaBean 属性通常遵循简单的方法命名规则，应用程序利用反射机制找到 JavaBean 提供的属性，查询或修改属性值，对 bean 对象进行读写操作；JavaBean 还可以对属性值的改变做出及时的响应。

2. 方法

JavaBean 中的方法就是对普通的 Java 方法进一步约束，其方法由 public 修饰，其他组件可自由调用，不同的方法有不同的约束，例如：bean 构造方法必须是无参的；读取 bean 的属性值方法为 getXXX()，返回类型不受限制；设置属性值的方法为 setXXX()，返回类型为 void，其中"XXX"表示属性名称，且首字母大写。

由于 JavaBean 本身是 Java 对象，调用对象的方法是与其交互作用的唯一途径。JavaBean 严格遵守面向对象的类设计逻辑，不让外部对象访问其任何私有字段。

3. 事件

事件为 JavaBean 组件提供了一种发送通知给其他组件的方法。在 AWT 事件模型中，一个事件源可以注册事件监听器对象。当事件源检测到发生了某种事件时，它将调用事件监听器对象中的一个适当的事件处理方法来处理这个事件。

7.1.2 JavaBean 作用范围

JavaBean 在 JSP 页面中使用时存在生命周期，根据其作用范围的不同，其生命周期长短也不一样，JavaBean 的作用范围分为：页面 page、请求 request、对话 session、应用 application 四种。

■ 页面 page

JavaBean 的页面 page 作用范围只在当前页面有效，跳出当前页面则无法使用当前页面的 JavaBean 对象。

■ 请求 request

JavaBean 的请求 request 对象作用范围可以将当前页面定义的 JavaBean 对象通过动作指令 jsp:forward 跳转到目标页面后，通过 request.getAttribute（JavaBean 对象的 id）获取已经存在的 JavaBean 对象。

■ 对话 session

JavaBean 的对话 session 作用范围可以将当前页面创建的 JavaBean 对象保存在 session 的会话中，通过 session.getAttribute（JavaBean 对象的 ID）获取已经存在的 JavaBean 对象。

session 会话对象一结束，JavaBean 对象立即消失。

- 应用 application

JavaBean 的应用程序 application 作用范围可以将当前页面设置的 JavaBean 对象保存在应用程序 application 的属性中，通过 application.getAttribute（JavaBean 对象的 ID）获取保存的 JavaBean 对象，只要当前应用程序一直处于运行状态，都可以随时获取到 JavaBean 的数据，一旦当前应用程序停止运行，JavaBean 的对象也会消失。

7.2 JavaBean 与动作指令应用

动作指令 jsp:useBean、jsp:setProperty 和 jsp:getProperty 等与 JavaBean 紧密相连，通过 jsp:useBean 动作指令创建 JavaBean 对象，并设置其作用范围；jsp:setProperty 动作指令设置 JavaBean 的属性值；而 jsp:getProperty 动作指令获取 JavaBean 中的属性值。本节通过案例说明 JavaBean 与这些动作指令之间的关系及应用。

案例描述

创建一个角色类（Actor），要求角色类是符合 JavaBean 要求的类，该类含有属性角色编号（number），角色名称（name）和角色描述（description）；在页面 actor.jsp 中创建四个不同作用域的角色对象，并且当前页面的最后通过 jsp:forward 动作指令跳转到目标页面 getActorBean.jsp，在该页面分别显示 request 作用域的角色 JavaBean 对象值、会话 session 作用域的角色 JavaBean 对象值、应用 application 作用域的角色 JavaBean 对象的值。如图 7-1 所示。

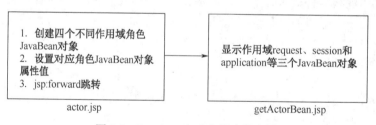

图 7-1　JavaBean 与动作指令的应用

图 7-1 中，显示了当前案例需要的两个页面 actor.jsp 和 getActorBean.jsp，页面 actor.jsp 实现对不同作用域的角色 JavaBean 进行初始化，并且对其属性进行赋值，赋值结束后，通过 jsp:forward 跳转到 getActorBean.jsp 页面；getActorBean.jsp 页面实现接收来自 actor.jsp 页面角色对象数据，其作用域分别是 request、session 和 application 等。

案例分析

根据案例描述，须完成的任务有：

- 开发 JavaBean 特征的角色类 Actor，其属性包括角色编号、名称及描述等；具体设计见表 7-1。

表 7-1　Actor 类

类名称	Actor	中文名称	角色类	类　型	实体类
类描述	提供角色基本特性描述，具备 JavaBean 定义要求				
属性定义					
序　号	属性名	访问权限及修饰词	类　型	说　明	
1	number	private	String	角色编号	
2	name	private	String	角色名称	
3	description	private	String	角色描述	
方法定义					
序　号	方法名	参　数	返回类型	功能说明	
1	getNumber	无	String	获取角色编号属性值	
2	setNumber	String number：角色编号	void	设置角色编号属性值	
3	getName	无	String	获取角色名称属性值	
4	setName	String name：角色名称	void	设置角色名称属性值	
5	getDescription	无	String	获取角色描述属性值	
6	setDescription	String description：角色描述	void	设置角色描述属性值	

- 开发 actor.jsp 页面，并在页面中创建四个不同作用域的角色对象，实现 forward 跳转；
- 开发接收页面 getActorBean.jsp，在页面中接收 request、session 和 application 等不同作用域中的角色对象，并显示这些角色对象信息。

案例实现

根据案例分析，通过 3 个步骤实现：

1. 创建角色 Actor 类

根据角色类表 7-1 的设计，角色类 Actor 的实现代码如下：

```java
package chapter7;
/**
 * 角色JavaBean类
 * Created by wph-pc on 2018/12/10.
 */
public class Actor {
    //角色编号
    private String number;
    //角色名称
    private String name;
    //角色描述
    private String description;
    public String getNumber() {
        return number;
    }
    public void setNumber(String number) {
        this.number = number;
    }
    public String getName() {
        return name;
    }
```

```java
    }
    public void setName(String name) {
        this.name = name;
    }
    public String getDescription() {
        return description;
    }
    public void setDescription(String description) {
        this.description = description;
    }
}
```

2. actor.jsp 页面

actor.jsp 页面实现了对角色类 Actor 进行 JavaBean 的初始化，并且通过 jsp:useBean、jsp:setProperty、jsp:getProperty 等动作实现角色对象的实例化及相关属性的赋值，具体实现代码如下：

```jsp
<%@ page import="chapter7.Actor" %><%--
  Created by IntelliJ IDEA.
  User: wph-pc
  Date: 2018/12/8
  Time: 14:57
  To change this template use File | Settings | File Templates.
--%>
<%@ page contentType="text/html;charset=UTF-8" language="java" %>
<%@include file="../header/init_bootstrap.jsp"%>
<html>
<head>
    <title>JavaBean技术</title>
</head>
<body class="container">
<h2>第7章：JavaBean案例</h2>
<hr>
<!--page作用域-->
<jsp:useBean id="actor" class="chapter7.Actor" scope="page"></jsp:useBean>
    <jsp:setProperty name="actor" property="number" value="001"></jsp:setProperty>
    <jsp:setProperty name="actor" property="name" value="系统管理员"></jsp:setProperty>
    <jsp:setProperty name="actor" property="description" value="JavaBean的page作用域"></jsp:setProperty>

<!--request作用域-->
<jsp:useBean id="actor1" class="chapter7.Actor" scope="request"></jsp:useBean>
    <jsp:setProperty name="actor1" property="number" value="002"></jsp:setProperty>
    <jsp:setProperty name="actor1" property="name" value="系统管理员2"></jsp:setProperty>
    <jsp:setProperty name="actor1" property="description" value="JavaBean的request作用域"></jsp:setProperty>
```

```
        <!--session作用域-->
        <jsp:useBean id="actor2" class="chapter7.Actor" scope="session"></jsp:
useBean>
        <jsp:setProperty name="actor2" property="number" value="003"></jsp:
setProperty>
        <jsp:setProperty name="actor2" property="name" value="系统管理员3"></jsp:
setProperty>
        <jsp:setProperty name="actor2" property="description" value="JavaBean
的session作用域"></jsp:setProperty>

        <!--application作用域-->
        <jsp:useBean id="actor3" class="chapter7.Actor" scope="application">
</jsp:useBean>
        <jsp:setProperty name="actor3" property="number" value="004"></jsp:
setProperty>
        <jsp:setProperty name="actor3" property="name" value="系统管理员4"></jsp:
setProperty>
        <jsp:setProperty name="actor3" property="description" value="JavaBean
的application作用域"></jsp:setProperty>

        <jsp:forward page="getActorBean.jsp"></jsp:forward>
    </body>
    </html>
```

3. getActorBean.jsp 页面

getActorBean.jsp 页面实现对不同作用域的角色对象进行呈现，主要是 request、session 和 application 等作用域，通过它们的方法 getAttribute 来获取存放的对象，具体实现代码如下：

```
<%@ page import="chapter7.Actor" %><%--
    Created by IntelliJ IDEA.
    User: wph-pc
    Date: 2018/12/12
    Time: 20:30
    To change this template use File | Settings | File Templates.
--%>
<%@ page contentType="text/html;charset=UTF-8" language="Java" %>
<html>
<head>
    <title>获取JavaBean对象</title>
</head>
<body>
<!--request作用域-->
    <%
        Actor obj=(Actor)request.getAttribute("actor1");
        out.print("<h2>=========JavaBean的request作用域=========</h2>");
        out.print("角色名称:"+obj.getName()+";角色描述:"+obj.getDescription()+
"<br>");
    %>
<!--session作用域-->
    <%
```

```
        Actor obj1=(Actor)session.getAttribute("actor2");
        out.print("<h2>=========JavaBean的session作用域========</h2>");
        out.print("角色名称:"+obj1.getName()+";角色描述:"+obj1.getDescription()+
"<br>");
    %>
    <!--application作用域-->
    <%
        Actor obj2=(Actor)application.getAttribute("actor3");
        out.print("<h2>=========JavaBean的application作用域========</h2>");
        out.print("角色名称:"+obj2.getName()+";角色描述:"+obj2.getDescription()+
"<br>");
    %>
    </body>
    </html>
```

运行结果

运行 actor.jsp 页面，页面不会停留在 actor.jsp，而是会直接跳转到 getActorBean.jsp 页面，如图 7-2 所示。

```
=========JavaBean的request作用域========
角色名称：系统管理员2;角色描述：JavaBean的request作用域

=========JavaBean的session作用域========
角色名称：系统管理员3;角色描述：JavaBean的session作用域

=========JavaBean的application作用域========
角色名称：系统管理员4;角色描述：JavaBean的application作用域
```

图 7-2　JavaBean 与动作指令案例运行结果

7.3　JSON 与 JavaBean 转换应用

JSP 页面与控制层进行数据交互时，可以采用 JSON 格式的 Ajax 技术，例如新增学生对象到数据库，通常需要将页面上的各种学生属性信息按照 JSON 格式要求进行编写，并与学生实体层的属性相对应，在 Java Web 后台（控制层或 Servlet）可以将传递过来的 JSON 格式的学生数据转换成学生 JavaBean 对象。

什么是 JSON 数据格式？JSON 是（JavaScript Object Notation，JS 对象简谱）一种轻量级的数据交互格式，由于其层次结构清晰，编写简单，广泛应用于网络数据交互。JSON 对象格式描述定义：

```
        var obj={属性名1:属性值1，属性名2:属性值2，…};
```

JSON 对象数据格式表示中，"{}"大括号表示对象，里面由属性组成，属性按照"属性名:属性值"格式编写，属性与属性之间用"，"隔开。

JSON 对象中,集合采用"[]"表示,可以与 JSON 中的"{}"对象格式结合使用。例如:
```
var obj=[{},{},{}];
```
或者:
```
var obj={属性名:[集合]};
```

Ajax 即"Asynchronous Javascript And XML"(异步 JavaScript 和 XML),是指一种异步数据交互的网页开发技术,有时也称局部刷新技术。采用这种技术后,页面不再频繁地刷新或跳转,提升了用户 UI 体验感。本节 Ajax 采用了 JQuery 中的 Ajax。

案例描述

采用 JSON 格式的 Ajax 数据交互技术,将页面采集到的角色信息提交到另外一个 JSP 页面,该页面将接收到具有 JSON 格式的角色信息转换成角色 JavaBean 对象,并返回到提交页面。角色信息采集页面如图 7-3 所示。

图 7-3 角色信息采集页面

图 7-3 中,分采集角色编号、角色名称和角色描述三个不同角色属性,填写完毕后,单击"提交数据"按钮,通过将角色的三个属性按照 JSON 格式及 JavaBean 属性名称组成角色 JSON 对象,利用 Ajax 提交到处理页面,并等待执行结果。

案例分析

根据案例描述要求,须解决以下问题:

- 创建采集角色编号、角色名称及角色描述的页面,定义该页面名称为 actorCommit.jsp;
- 提交采集页面的数据,需要使用 JSON 格式,即 JSON 对象的属性名称要与角色的 JavaBean 属性名称保持一致;
- 需要有接收来自 actorCommit.jsp 提交数据处理的页面,定义该页面名称为 getBean.jsp;该页面需要将获取的角色数据转换成 Actor 的 JavaBean 对象,并将获取到的 JavaBean 对象以 JSON 格式返回到 actorCommit.jsp 页面。

角色 JSON 与 JavaBean 转换示意图如图 7-4 所示。

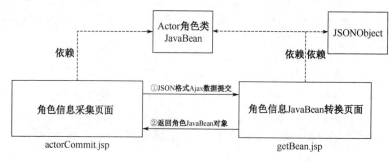

图 7-4 角色 JSON 与 JavaBean 转换示意图

图 7-4 中，角色信息采集页面 actorCommit.jsp 利用 HTML 标签通过表单的形式采集角色编号、角色名称及角色描述信息。表单中特别需要注意的是，需要将角色编号、角色名称及角色描述属性的"name"属性值与 Actor 角色类中对应的属性名设置为一致。将采集到的角色信息形成 JSON 对象，利用 JQuery 中的 Ajax 技术提交到 getBean.jsp 页面，getBean.jsp 页面接收到数据后，通过 JSONObject 对象转换成角色 Actor 的 JavaBean 对象，并将获取到的角色对象以 JSON 格式形式返回到原调用页面 actorCommit.jsp。

在此方案中，前端的 Ajax 的 JSON 格式数据交互需要使用 JQuery，版本建议在 1.10 以上，页面 BootStrap 也需要依赖 JQuery。另外，后台需要使用 JSONObject 类。

案例实现

1. 角色信息采集页面 actorCommit.jsp

角色信息采集页面采用 BootStrap 构建，如果读者还未学习，可使用常规的 HTML 构建即可，具体实现如下：

```jsp
<%@ page import="chapter7.Actor" %><%--
  Created by IntelliJ IDEA.
  User: wph-pc
  Date: 2018/12/8
  Time: 14:57
  To change this template use File | Settings | File Templates.
--%>
<%@ page contentType="text/html;charset=UTF-8" language="java" %>
<%@include file="../header/init_bootstrap.jsp"%>
<html>
<head>
    <title>JavaBean技术</title>
</head>
<body class="container">
<h2>第7章：角色JavaBean数据提交</h2>
<hr>
<div class="panel panel-primary">
    <div class="panel-heading panel-primary">
       角色JavaBean案例
    </div>
    <div class="panel-body">
        <div class="row">
            <div class="col-md-12 col-sm-12 col-lg-12 col-xs-12">
                <form method="post" id="ff">
```

```html
                    <div class="form-group">
                        <label for="txtNumber">角色编号</label>
                        <input type="text" class="form-control" name="number" placeholder="请输入角色编号" id="txtNumber"/>
                    </div>
                    <div class="form-group">
                        <label for="txtName">角色名称</label>
                        <input type="text" class="form-control" name="name" id="txtName"/>
                    </div>
                    <div class="form-group">
                        <label for="txtDescription">角色描述</label>
                        <input type="text" class="form-control" name="description" id="txtDescription"/>
                    </div>
                    <button type="button" id="btnCommit" class="btn btn-primary">提交数据</button>
                </form>
            </div>
        </div>
    </div>
</div>
<script>
    $(function () {
        $("#btnCommit").click(function () {
//获取表单ff中的JSON格式数据
            var obj={number:$("#txtNumber").val(),name:$("#txtName").val(),description:$("#txtDescription").val()};
doData("chapter7/getBean.jsp",obj,function (result) {
                alert("服务器返回角色信息=>编号: "+
                result.number+"; 名称: "+
                result.name+"; 描述: "+result.description);
            },true);
        });
    });
</script>
</body>
</html>
```

以上实现过程中，使用了 JavaScript 函数 doData()，利用 JQuery 的 Ajax 实现了异步数据交互，定义如下：

```
    /*********************************************
     * Ajax数据处理
    *url:请求地址
     * params:参数对象
     * callback:回调函数
     * mask:数据交互是否使用遮罩，true表示使用,不填或false表示不使用
     *********************************************/
    function doData(url,params,callback,mask) {
        url=getRootPath()+"/"+url;//项目的根地址
        $.Ajax({
            type : 'post',
```

```javascript
            url : url,
            dataType: 'json',
            contentType: "application/json; charset=utf-8",
            cache:true,
            data: JSON.stringify(params),
            beforeSend: function(){
                if (mask!=undefined && mask==true && $("#mask").length>0)
                {
                    $("#mask").css("height",$(document).height());
                    $("#mask").css("width",$(document).width());
                    $("#mask img").css("padding-top",window.innerHeight*0.45);
                    $("#mask").show();
                }
            },
            complete:function () {
                if (mask!=undefined && mask==true && $("#mask").length>0)
                    $("#mask").hide();
            },
            success: function (data) {
                if (mask!=undefined && mask==true && $("#mask").length>0)
                    $("#mask").hide();
                if (callback) callback(data);
            },
            error : function(arg0,arg1,arg2) {
                switch(arg0.status)
                {
                    case 200:
                        alert("服务器已经接收到您的请求,但无法做出正确的响应,请联系管理员进行处理,问题发生地址: " + url);
                        break;
                    case 404:
                        alert("当前操作的资源不存在,请联系管理员!");
                        break;
                    case 500:
                        alert("程序内容处理错误: 500,内部符号"+url);
                        break;
                    default:
                        alert("数据处理错误,错误代码: "+arg0.status);
                        break;
                }
            }
        });
    }
```

2. 角色 Actor 类

角色 Actor 类是标准的 JavaBean 类,前面的内容有所涉及,此处不再讲述。

3. getBean.jsp 页面

getBean.jsp 页面获取到来自于 actorCommit.jsp 页面的信息后,将角色信息按照角色 Actor 的 JavaBean 标准转成角色对象,转换中,利用 JSONObject 类,转换完成后,再利用

JSON.toJSONString()将角色对象转成 JSON 字符串,并返回 actorCommit.jsp 页面。特别提示,用内置对象 response 返回 JSON 格式需要进行以下设置:

```
response.setHeader("Content-Type", "application/json;charset=UTF-8");
```

页面具体实现代码如下:

```jsp
<%@ page import="util.JSONTool" %>
<%@ page import="com.alibaba.fastjson.JSONObject" %>
<%@ page import="com.alibaba.fastjson.JSON" %>
<%@ page import="chapter7.Actor" %><%--
  Created by IntelliJ IDEA.
  User: wph-pc
  Date: 2018/12/13
  Time: 8:03
  To change this template use File | Settings | File Templates.
--%>
<%@ page contentType="text/html;charset=UTF-8" language="java" %>
<html>
<head>
    <title>获取角色Bean数据</title>
</head>
<body>
  <%
    //设置返回的格式
    response.setHeader("Content-Type", "application/json;charset=UTF-8");
    //将客户端传递的参数转换成JSON格式字符串
    String json= JSONTool.GetPostData(request.getInputStream(), request.getContentLength(), "utf-8");
    Actor actor=JSONObject.parseObject(json, Actor.class);
    response.getWriter().write(JSON.toJSONString(actor));
    response.getWriter().close();
  %>
</body>
</html>
```

以上代码中,JSONTool.GetPostData(request.getInputStream(), request.getContentLength(), "utf-8")实现从客户端请求对象 request 中获取传递过来的角色对象 JSON 格式字符串,本书将这个方法放在 JSONTool 自定义类中统一管理,该方法定义如下:

```java
/*
 *获取输入流in对象的字符串值
 * @param in:客户端输入流对象
 * @param size:输入流大小
 * @param charset:字符集名称
 * @return 返回字符流字符串内容
 **/
public static String GetPostData(InputStream in, int size, String charset) {
    if (in != null && size > 0) {
        byte[] buf = new byte[size];
        try {
            //读取输入流数据到buf字节数组
            in.read(buf);
            //如果字符集为空,默认为"utf-8"
            if (charset == null || charset.length() == 0)
```

```
                return new String(buf,"utf-8");
            else {
                return new String(buf,charset);
            }
        } catch (IOException e) {
            e.printStackTrace();
        }
    }
    return null;
}
```

运行结果

运行 actorCommit.jsp 页面,其行页面如图 7-3 所示,填写角色编号为:001,角色名称为:超级用户,角色描述为:系统管理员,单击"提交数据"按钮,弹出如图 7-5 所示的提示框。

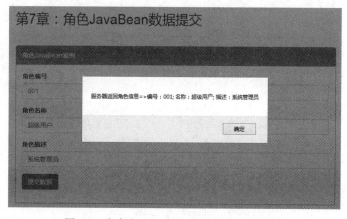

图 7-5 角色与 JavaBean 数据提交运行结果

图 7-5 中,页面不需要跳转就可以获得来自服务器端 getBean.jsp 页面的 JavaBean 数据响应。

习题

按照 JavaBean 技术要求,定义书籍的 JavaBean 类,书籍的属性包括:ISBN、书籍名称、作者、出版社、出版日期、版本等。项目操作具体要求如下:

1. 按照书籍提供的属性,在服务器端定义书籍(Book)JavaBean 类;
2. 提供书籍信息采集的 JSP 页面,采用 Ajax 与 JSON 格式实现数据交互;
3. 提供书籍信息显示 JSP 页面,将存放在服务器上的书籍 JavaBean 对象转换成 JSON 格式后显示在页面上。

第 8 章

Servlet 技术

JSP 页面编程时，经常需要将 Java 代码、HTML 标签、CSS 样式、JS 脚本放在一个 JSP 页面中，整个页面看起来比较杂，管理也很麻烦。最好的处理方式是将 JSP 页面中的 Java 代码与 HTML 静态页面内容进行分离，Servlet 技术就实现了此要求。只要将 JSP 页面中的 Java 部分代码全部移植到 Servlet 类中，原有的功能不受任何影响。本章主要讲解 Servlet 的相关概念、Servle 的创建及使用，以及 Servlet 在项目中的应用。

本章任务

（1）Servlet 的相关概念；
（2）Servlet 的创建及使用；
（3）Servlet 在项目中的应用。

重点内容

（1）掌握 Servlet 的创建及使用；
（2）掌握 Servlet 在项目中的应用。

难点内容

Servlet 在项目中的应用。

8.1　Servlet 相关知识

Java 中的 Servlet 是指运行在 Web 服务器上的程序，它可以：
- 获取来自页面上的各种信息；
- 将服务器上的各种信息发送到客户端；
- 动态创建网页；

- 将 JSP 页面中的页面标签与 Java 代码完全分离。

Servlet 是 Java 中的类，可以在类中调用 JSP 的各种内置对象，它是 Web 页面与后台 Java 程序之间的中间件，通过它可以将用户页面上的各种信息传给后台处理程序，后台处理程序的处理结果也可以通过 Servlet 响应到客户端页面。Servlet 在项目处理中的作用如图 8-1 所示。

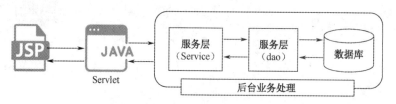

图 8-1　Servlet 在项目处理中的作用

从图 8-1 中可以看出，Servlet 在项目中实现对 JSP 页面的数据进行采集，传递给服务层处理，如果需要数据库支持，服务层将请求数据访问层 dao，数据访问层 dao 对数据库进行操作，操作结束后将结果返回给数据访问层，数据访问层再将结果发送到服务层 Service，服务层 Service 将执行结果告知 Servlet，最后由 Servlet 传递给 JSP 页面，这样就完整地完成了一次"双工"的操作。

8.1.1　Servlet 相关类

Servlet 指的是 HttpServlet 类，它是抽象类，具有用来专门处理 get 数据请求的 doGet 方法和 post 数据请求的 doPost 方法。在实际使用过程中，可以对 doGet 方法或 doPost 方法进行重写操作。

doGet 方法语法定义格式：

```
    protected void doGet(HttpServletRequest req, HttpServletResponse resp)
throws ServletException, IOException {
        String protocol = req.getProtocol();
        String msg = lStrings.getString("http.method_get_not_supported");
        if(protocol.endsWith("1.1")) {
            resp.sendError(405, msg);
        } else {
            resp.sendError(400, msg);
        }
    }
```

doPost 方法语法定义格式：

```
    protected void doPost(HttpServletRequest req, HttpServletResponse resp)
throws ServletException, IOException {
        String protocol = req.getProtocol();
        String msg = lStrings.getString("http.method_post_not_supported");
        if(protocol.endsWith("1.1")) {
          resp.sendError(405, msg);
        } else {
          resp.sendError(400, msg);
        }
    }
```

Servlet 存在生命周期，在创建 Servlet 对象时，会调用其方法 init，在 Servlet 运行期间，调用其方法 service，Servlet 销毁时调用其方法 destory，如图 8-2 所示。

图 8-2　Servlet 生命周期

图 8-2 中，Servlet 初始化只能调用一次，在对象创建时调用；service 方法在响应客户端请求时通过 get 或 post 方法调用，并将请求执行结果返回给客户端；destory 方法在 Servlet 对象销毁时调用，由 JVM 进行垃圾回收，只能调用一次，即生命周期结束时调用。Servlet 类描述见表 8-1。

表 8-1　Servlet 类描述

类名称	Servlet	中文名称	服务小应用程序	类　型	接口
类描述	提供 Servlet 功能操作的定义				
序　号	方法名	参　数	返回类型	功能说明	
1	init	ServletConfig var1：配置参数	void	Servlet 初始化	
2	getServletConfig	无	ServletConfig	获取 Servlet 配置对象	
3	service	ServletRequest var1：request 请求对象；ServletResponse var2：response 响应对象	void	响应客户端请求	
4	getServletInfo	无	String	获取 Servlet 相关信息	
5	destroy	无	void	Servlet 对象销毁	

8.1.2　Servlet 类定义方式

开发 Servlet 有三种方式：

■　实现 Serlvet 接口

语法格式：

```
package chapter8;
import javax.Servlet.*;
import java.io.IOException;
/**
 * Created by wph-pc on 2018/11/4.
 */
public class ServletMethod1 implements Servlet {
    @Override
    public void init(ServletConfig ServletConfig) throws ServletException {
```

```
        }
        @Override
        public ServletConfig getServletConfig() {
            return null;
        }
        @Override
        public void service(ServletRequest ServletRequest, ServletResponse
ServletResponse) throws ServletException, IOException {

        }
        @Override
        public String getServletInfo() {
            return null;
        }
        @Override
        public void destroy() {

        }
    }
```

直接实现 Servlet 接口需要在子类中实现接口中的所有方法，比较麻烦，如果只使用 service 方法，可使用第二种方式：通过继承 GenericServlet 类实现。

■ 继承 GenericServlet

语法格式：

```
package chapter8;
import javax.Servlet.GenericServlet;
import javax.Servlet.ServletException;
import javax.Servlet.ServletRequest;
import javax.Servlet.ServletResponse;
import java.io.IOException;

/**
 * Created by wph-pc on 2018/11/4.
 */
public class ServletMethod2 extends GenericServlet {
    @Override
    public void service(ServletRequest ServletRequest, ServletResponse
ServletResponse) throws ServletException, IOException {
    }
}
```

通过继承 GenericServlet 类的方式更加简单，只要实现父类中 service 接口即可。GenericServlet 抽象类中实现了 Servlet 的其他接口。

■ 继承 HttpServlet

语法格式：

```
package chapter8;
import javax.Servlet.http.HttpServlet;

/**
 * Created by wph-pc on 2018/11/4.
 */
```

```
public class ServletMethod3 extends HttpServlet {
}
```

通过直接继承 HttpServlet 的方式更加简单，HttpServlet 父类已全部实现了接口 Servlet，子类不需要实现父类中的任何方法，但在实际使用 HttpServlet 继承方式时，一般子类都需要重写父类中的 doGet 方法或 doPost 方法。如果采用 get 数据交互方式，则重写 doGet 方法，而采用 post 数据交互方式则需要重写 doPost 方法。

8.2 基于 Servlet 用户登录

Servlet 自定义业务类的创建及使用需要三个步骤，如图 8-3 所示。

图 8-3　Servlet 创建与应用步骤

图 8-3 中可以看出，Servlet 类的定义及对象创建需要三个步骤：
- 首先，根据三种不同 Servlet 定义方式，选择一种定义 Servlet 类；
- 其次，Servlet 对象的创建不需要通过 new 实例化，而是通过 Web 项目中的 web.xml 文件进行配置的；
- 最后，配置客户端请求的映射地址，同样在 web.xml 文件中的 Servlet-mapping 节点，其下有子节点 Servlet-name 和 Servlet-url，分别表示 Servlet 对象名称和客户端映射地址。请注意，这里的 Servlet-name 值要与 Servlet 节点中的 Servlet-name 值一致，否则映射的地址无法找到 Servlet 处理程序。

Servlet 中 web.xml 的配置参考格式：

```
<Servlet>
    <Servlet-name>Servlet对象名称</Servlet-name>
    <Servlet-class>Servlet类名称，含完整包名</Servlet-class>
</Servlet>
<Servlet-mapping>
    <Servlet-name>Servlet对象名称</Servlet-name>
    <url-pattern>请求地址，自定义</url-pattern>
</Servlet-mapping>
```

案例描述

用户身份验证采用 JSP 页面替代身份验证的后台功能，在 loginCheck.jsp 页面中的身份验证功能中使用 Servlet 技术，其他功能不变。

案例分析

用户身份验证中，需要对验证身份的页面 loginCheck.jsp 采用 Servlet 技术替换，定义 DoLogin 类，从 HttpServlet 类继承过来，重写 doPost 方法。

案例实现

1. 定义 DoLogin 类

```java
package chapter8;
import business.entity.User;
import javax.Servlet.ServletContext;
import javax.Servlet.ServletException;
import javax.Servlet.http.Cookie;
import javax.Servlet.http.HttpServlet;
import javax.Servlet.http.HttpServletRequest;
import javax.Servlet.http.HttpServletResponse;
import java.io.IOException;
import java.io.PrintWriter;

/**
 * Created by wph-pc on 2018/10/30.
 */
public class DoLogin extends HttpServlet {
    private String showErrorTips(String errMsg)
    {
        //错误提示信息
        return "<div class='well'>"+errMsg+",请重新<a href='chapter6/comprehensivecase/login.jsp'>登录</a></div>";
    }
    @Override
    public void doGet(HttpServletRequest request,
                      HttpServletResponse response)
        throws ServletException, IOException {
        doPost(request,response);
    }
    @Override
    public void doPost(HttpServletRequest request,
                       HttpServletResponse response)
        throws ServletException, IOException {
        //设置客户端字符编码
        request.setCharacterEncoding("utf-8");
        response.setContentType("text/html;charset=utf-8");
        //创建用户对象
        User user=new User();
        /*获取客户端账号、密码及是否保存密码标记*/
        user.setNumber(request.getParameter("number"));
        user.setPassword(request.getParameter("password"));
        String remember = request.getParameter("passcookies");
        PrintWriter out=response.getWriter();
        /*验证码验证*/
        if (request.getSession().getAttribute("code")==null ||
```

```java
                request.getSession().getAttribute("code") instanceof String ==false)
                {
                    out.print(showErrorTips("系统没有获取到验证码信息"));
                    return;
                }
                else
                {
                    String code=request.getSession().getAttribute("code").toString();
                    if (!code.equals(request.getParameter("code")))
                    {
                        out.print(showErrorTips("验证码错误"));
                        return;
                    }
                }
        /*验证账号与密码*/
            if (!"admin".equals(user.getNumber()) || !"admin".equals(user.getPassword())) {
                out.print("账号"+user.getNumber()+";密码: "+user.getPassword());
                out.print(showErrorTips("账号或密码错误"));
                return;
            } else {
                //将用户信息写入session中
                request.getSession().setAttribute("user",user);
                if (remember != null) {
                    Cookie c1 = new Cookie("number", user.getNumber());
                    Cookie c2 = new Cookie("password", user.getPassword());
                    //设置加密传输
                    c2.setSecure(true);
                    /*设置cookie的时长为15天,以秒为单位*/
                    c1.setMaxAge(15*24*60*60);
                    c2.setMaxAge(15*24*60*60);
                    /*将设置的cookie添加到response带回到客户端*/
                    response.addCookie(c1);
                    response.addCookie(c2);

                    out.println("<script>toastr.info('登录成功,并记住了账号与密码');</script>");
                } else {
                    out.println("<script>toastr.info('登录成功,但没有记住账号与密码');</script>");
                }
                //获取全局
                ServletContext application=getServletContext();
                /*更新访问数量*/
                if (application.getAttribute("count")==null || application.getAttribute("count") instanceof Integer==false)
                    application.setAttribute("count",1);
                else if (request.getSession().isNew())
                    application.setAttribute("count",(Integer)application.
```

```
getAttribute("count")+1);
                //跳转主页面index.jsp
                response.sendRedirect("chapter6/comprehensivecase/index.jsp");
        }
    }
}
```

通过实现 DoLogin 类的代码可以看出，Servlet 类中完全可以像 JSP 页面中一样正常使用内置对象，实现了页面请求 Java 后台功能，以及将 Java 后台功能的执行结果返回到页面。

2. 配置 web.xml 中的 Servlet 节点

- 配置 DoLogin 对象节点

```
<Servlet>
    <Servlet-name>checklogin</Servlet-name>
    <Servlet-class>chapter8.DoLogin</Servlet-class>
</Servlet>
```

配置节点中，Servlet-name 的名称由读者自定义，Servlet-class 节点必须是完整的类名称，含命名空间。

- 配置访问路径节点

```
<Servlet-mapping>
    <Servlet-name>checklogin</Servlet-name>
    <url-pattern>/logincheck</url-pattern>
</Servlet-mapping>
```

节点 Servlet-name 必须与 Servlet 节点中的 Servlet-name 子节点名称相同，url-pattern 节点是客户端请求路径，可以自定义。

3. 更改登录页面验证请求路径

将登录页面 login.jsp 中 form 表单的 action 属性的验证请求路径改为 "/logincheck" 即可。

运行结果

用户身份验证时，首先运行页面 login.jsp，结果如图 8-4 所示。

图 8-4 用户登录页面

单击图 8-4 中的 "登录" 按钮后，通过 "/logincheck" 登录验证，验证通过后，将跳转至 index.jsp 页面，结果如图 8-5 所示。

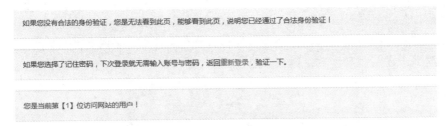

图 8-5　index.jsp 页面

8.3　基于 MVC 的三层架构用户管理

到目前为止，已经学习完 Servlet 的使用规则，通过用户登录身份验证实现了 Servlet 的初步应用。本节通过 Servlet 技术实现 MVC 框架，并与项目中的三层架构联合使用，实现构建 Web 项目开发框架。

MVC 是一种使用 MVC（Model View Controller 模型-视图-控制器）设计创建 Web 应用程序的模式：Model（模型）表示应用程序核心（比如数据库记录列表）；View（视图）显示数据（数据库记录）；Controller（控制器）处理输入（写入数据库记录）。

三层架构分为表示层、业务层与数据访问层，MVC 实现了表示层的功能，三层之间的关系是：当数据从表示层发送请求后，请求的指令传递给业务层，如果需要进行数据库操作，业务层再将指令发送给数据访问层，数据访问层处理结束后，再将指令反向传递给业务层，业务层再将执行结果传给表示层。

MVC 作为三层架构中的表示层，它们之间具体的关系如图 8-6 所示。

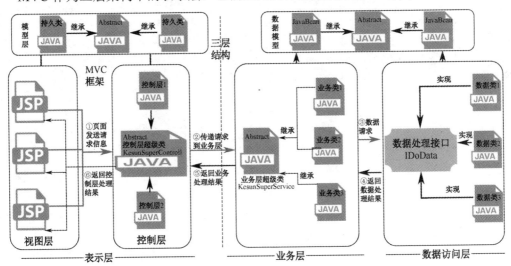

图 8-6　三层架构与 MVC 项目关系

1．表示层设计

图 8-7 中表示层采用了 MVC 模式，充分利用 Servlet 实现控制层功能。MVC 框架实现了表示层，视图层主要由 JSP 页面组成，控制层由 Servlet 实现，模型层由 JavaBean 实体

模型类组成。在视图层中，项目设计了一个 KesunSuperController 控制层超级类，它是一个 Servlet 类，实现了常规的实体数据的增、删、改、查等操作，各种控制层类都要从该超级控制层类继承，在后面案例中将详细介绍该类。

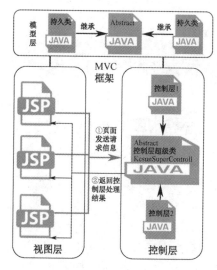

图 8-7　表示层设计

（1）控制层类。

该层设计了 KesunSuperController 控制层超级类，将控制层常规功能进行了统一定义及实现，为项目中其他模块提供了标准，它是一个 Servlet 类。KesunSuperController 控制层超级类见表 8-2。

表 8-2　KesunSuperController 控制层超级类

类 名 称	KesunSuperController	中文名称	控制层超级类	类　型	抽象类
类 描 述	提供控制层各模块的 CRUD 统一操作				
成员变量					
序　号	变 量 名	访问权限	类　型	说　明	
1	bll	private	KesunSuper-Service	业务层访问对象	
方　　法					
序　号	方 法 名	参　数	返回类型	功能说明	
1	createModel	HttpServletRequest request：客户端请求对象	AbsObject	抽象方法，根据客户端请求内容，产生 Model 数据实体	
2	createCondition	HttpServletRequest request：客户端请求对象	Map<String, Object>	抽象方法，根据客户端请求参数，转换成 Map 对象格式	
3	judgeService	无	KesunReturn	私有方法，判断业务对象是否有效	
4	add	HttpServletRequest request：客户端请求对象	KesunReturn	实现对象的新增功能	
5	edit	HttpServletRequest request：客户端请求对象	KesunReturn	实现对象的修改功能	

续表

		方 法		
序 号	方法名	参 数	返回类型	功能说明
6	del	HttpServletRequest request：客户端请求对象	KesunReturn	实现对象的删除功能
7	changeStatus	HttpServletRequest request：客户端请求对象	KesunReturn	实现对象的状态变更功能
8	getMe	HttpServletRequest request：客户端请求对象	KesunReturn	实现单个对象的查找功能
9	find	HttpServletRequest request：客户端请求对象	KesunReturn	实现对象的查找功能，以对象列表返回
10	findForMap	HttpServletRequest request：客户端请求对象	KesunReturn	实现对象的查找功能，以二维表格形式返回

（2）数据模型类。

① AbsObject 超级抽象模型类。AbsObject 是数据模型层中所有模型的超级抽象类，各模型层必须从该类继承，AbsObject 类定义见表 8-3。

表 8-3 AbsObject 类

类名称	AbsObject	中文名称	实体模型超级类	类 型	抽象类
类描述	提供数据模型实体层对象共同属性特征定义				
成员变量					
序 号	变量名	访问权限	类 型	说 明	
1	id	private	String	对象 ID 属性	
2	name	private	String	对象名称属性	
3	createDate	private	String	对象创建日期属性	
4	description	private	String	对象描述说明属性	
5	status	private	String	对象状态属性	
方 法					
序 号	方法名	参 数	返回类型	功能说明	
1	setId	String id：对象 id	void	设置对象 ID 属性值	
2	getId	无	String	获取对象 ID 属性值	
3	setName	String name：对象名称	void	设置对象名称	
4	getName	无	String	获取对象名称	
5	setCreateDate	Date createDate：对象创建日期	void	设置对象创建日期	
6	getCreateDate	无	Date	获取对象创建日期	
7	setDescription	String description：对象描述	void	设置对象描述信息	
8	getDescription	无	String	获取对象描述信息	
9	setStatus	String status：对象状态	void	设置对象状态信息	
10	getStatus	无	String	获取对象状态信息	

② KesunReturn 返回模型类。为了更加方便地描述各功能方法返回结果的多样化，对返回结果的表示形式定义了一个返回类 KesunReturn，它包括状态码 code 属性、message 消息属性和 obj 返回结果属性值，具体定义见表 8-4。

表 8-4　KesunReturn 类

类 名 称	KesunReturn	中文名称	返回模型类	类　　型	实体类	父类	无
类 描 述	对功能执行结果进行定义						
成员变量							
序　　号	变量名	访问权限	类　　型		说　　明		
1	code	private	String		状态码		
2	message	private	String		消息		
3	obj	private	Object		对象		
方　　法							
序　　号	方法名	参　　数	返回类型		功能说明		
1	setCode	String code：状态码	void		设置返回状态码		
2	getCode	无	String		获取返回状态码		
3	setMessage	String message：消息	void		设置返回消息		
4	getMessage	无	String		获取返回消息		
5	setObj	Object obj：对象	void		设置返回对象		
6	getObj	无	Object		获取返回对象		

（3）视图层。

视图层由 JSP 页面或 HTML 页面组成，模型层支持页面的数据呈现。

2. 业务层设计

业务层是项目中的核心功能模块，它向控制层提供了用户操作 UI 的所有功能，同时也是获取数据访问层功能的重要媒介。图 8-8 展示了业务层的设计理念。

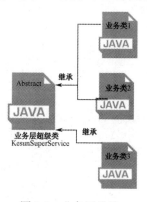

图 8-8　业务层设计

图 8-8 中，所有的业务层公共功能全部封装在业务层超级类 KesunSuperService 中，该类实现了常规的 CRUD 操作，所有的业务层类都必须从该类中继承。KesunSuperService 业务层超级类的定义见表 8-5。

表 8-5　KesunSuperService 业务层超级类

类 名 称	KesunSuperService	中文名称	业务层超级类	类　型	抽象类	父类	无
类 描 述	提供业务层公共功能操作						
成员变量							
序　号	变量名	访问权限		类　型	说　　明		
1	model	private		AbsObject	业务层操作实体模型对象		
方　　法							
序　号	方法名	参　　数		返回类型	功能说明		
1	getModel	无		AbsObject	获取数据实体		
2	setModel	AbsObject model：操作数据实体		void	设置业务层操作数据实体		
3	getDAO	无		IDoData	获取数据访问层对象		
4	add	无		KesunReturn	新增操作		
5	edit	无		KesunReturn	修改操作		
6	del	无		KesunReturn	删除操作		
7	find	Map values：查询条件		KesunReturn	对象查找操作，以对象列表返回		
8	findForMap	Map values：查询条件		KesunReturn	查找操作，以 Map 结构形式的列表返回		
9	getMe	无		KesunReturn	获取单个对象		
10	changeStatus	无		KesunReturn	改变对象状态属性		

3．数据访问层设计

数据访问层设计如图 8-9 所示，数据访问层实现数据库功能处理，实现对 CRUD 的操作，IDoData 接口中将 CRUD 的常规操作进行约束性定义，所有的数据访问层必须实现对 IDoData 接口的实现或继承。IDoData 接口定义见表 8-6。

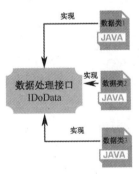

图 8-9　数据访问层设计

表 8-6　IDoData 类

类 名 称	IDoData	中文名称	数据访问层数据处理接口	类　型	接口	父类	无
类 描 述	提供数据访问层公共功能接口定义						
方　　法							
序　号	方法名	参　　数		返回类型	功能说明		
1	add	AbsObject obj：新增的对象		int	将 obj 参数对象新增到数据库中，返回值大于 0 表示成功，其他表示失败		

续表

序号	方法名	方法 参数	返回类型	功能说明
2	edit	AbsObject obj：修改的对象	int	对 obj 参数对象进行修改，返回值大于 0 表示成功，其他表示失败
3	del	AbsObject obj：删除的对象	int	将 obj 参数对象从数据库中删除，返回值大于 0 表示成功，其他表示失败
4	changeStatus	AbsObject obj：状态变更的对象	int	改变 obj 参数对象的状态，返回值大于 0 表示成功，其他表示失败
5	findResult	Map<String,Object> condition：查询条件	List<Map<String,Object>>	根据条件 condition 从数据库中查找数据，如果查询到数据，则返回 Map 结构数据，否则返回 null
6	find	Map<String,Object> condition：查询条件	List <AbsObject>	根据条件 condition 从数据库中查找数据，如果查询到数据，则返回 AbsObject 类型数据，否则返回 null
7	getMe	String id：查询对象的 ID 编号	AbsObject	根据参数 ID，从数据库中查找，如果查找到，返回目标对象，否则返回 null

案例描述

利用 Servlet 技术，实现 MVC 机制，并按照项目三层架构实现对用户模块的管理，管理的内容包括：

- 用户信息新增；
- 用户信息修改；
- 用户信息删除；
- 用户状态变更；
- 用户信息查找；
- 用户信息单个对象查找。

实现要求：

- 到目前为止因未学习数据库编程知识，因此数据访问层只要模拟功能实现即可；
- 业务层必须实现对数据访问层全部功能的调用；
- 表示层采用基于 Servlet 的 MVC 机制，MVC 中，控制层要求实现对业务层所有功能的调用，视图层 JSP 只要求实现用户信息新增页面即可；Model 数据模型层要求实现用户的 JavaBean 类。

案例分析

根据案例描述，需要采用 MVC 机制与三层架构技术，根据设计理念，将用户管理模块设计成如图 8-10 所示的结构。

图 8-10 中，表示层采用了 MVC 机制，需要使用 user.jsp 视图层，控制层需要 UserController 控制层类，它从 KesunSuperController 控制层超级类继承而来；数据模型采用 User 类表示，并且从 AbsObject 超级实体类继承而来。业务层中，由 UserServiceImpl 用户管理业务类实现各种用户管理功能，它继承了业务层超级类 KesunSuperService；数据访问层的功能由 UserDaoImpl 类实现。

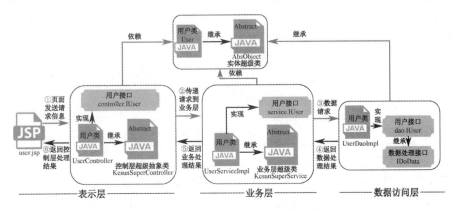

图 8-10 用户管理 MVC 与三层架构设计

1. 用户管理表示层设计

图 8-11 中,用户管理表示层对 MVC 设计理念进行了应用,UserController 用户控制层类从 KesunSuperController 控制层超级类继承而来,并实现了接口 IUser,接口是作为扩展使用,方便今后对功能进行扩充,案例中该接口没有具体的实际内容。

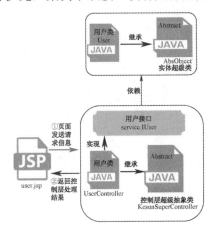

图 8-11 用户管理表示层设计

(1)控制层类。

用户控制层类 UserController 的定义见表 8-7。

表 8-7 UserController 类

类 名 称	UserController	中文名称	用户控制层类	类 型	实体类	父类	KesunSuperController
类 描 述	提供用户控制层的 CRUD 操作,从 KesunSuperController 控制层超级类继承,实现该层接口 IUser							
方 法							
序 号	方法名	参 数		返回类型			功能说明
1	UserController	无		UserController			构造方法,创建对象,并且实现对业务层对象参数的实例化
2	createCondition	HttpServletRequest request:客户端请求对象		Map<String, Object>			重写父类方法,根据客户端请求参数,转换成 Map 对象格式
3	createModel	HttpServletRequest request:客户端请求对象		AbsObject			重写父类方法,根据客户端请求内容,生产 Model 数据实体

续表

		方 法		
序 号	方法名	参 数	返回类型	功能说明
4	doPost	HttpServletRequest request：客户端请求对象；HttpServletResponse response：服务器端响应对象	void	实现用户对象 CRUD 各种操作

（2）User 模型类。

User 模型类用户表示用户的基本属性特征，从 AbsObject 继承而来，增加了 number 账号属性、nickName 昵称属性和 password 密码属性等，具体定义见表 8-8。

表 8-8　User 类

类名称	User	中文名称	用户模型类	类 型	实体类	父类	AbsObject
类描述	对用户基本属性进行定义。						
成员变量							
序 号	变量名	访问权限		类 型		说 明	
1	number	private		string		账号	
2	nickName	private		string		昵称	
3	password	private		string		密码	
方 法							
序 号	方法名	参 数		返回类型		功能说明	
1	setNumber	String number：账号		void		设置用户账号信息	
2	getNumber	无		string		获取用户账号	
3	setNickName	String nickName：昵称		void		设置用户昵称	
4	getNickName	无		string		获取用户昵称	
5	setPassword	String password：密码		void		设置用户密码	
6	getPassword	无		string		获取用户密码	

2. 用户管理业务层设计

用户管理业务层设计如图 8-12 所示，用户管理业务层设计中要求用户管理业务类 UserServiceImpl 实现 IUser 业务层接口，并且从业务层超级类 KesunSuperService 继承，业务层 IUser 接口作为扩展功能使用。UserServiceImpl 类定义见表 8-9。

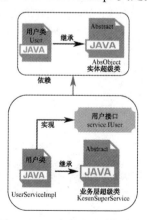

图 8-12　用户管理业务层设计

表 8-9 UserServiceImpl 类

类名称	UserServiceImpl	中文名称	用户业务类	类型	实体类	父类	KesunSuper-Service
类描述	提供对用户管理的各种功能操作，实现该层 IUser 接口，并从 KesunSuperService 业务层超级类继承						
方法							
序号	方法名	参数		返回类型			功能说明
1	UserServiceImpl	无		UserServiceImpl			构造方法，配置用户业务层实体对象
2	getDAO	无		IDoData			获取用户数据访问层对象

（1）用户管理数据访问层设计。

用户管理数据访问层设计如图 8-13 所示，用户数据访问层类 UserDaoImpl 实现该层接口 IUser，IUser 必须从 IDoData 继承，该层 IUser 接口仅从 IDoData 继承，没有其他内容，这里就不再展示其设计内容。UserDaoImpl 类的设计见表 8-10。

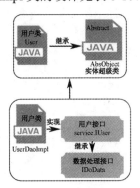

图 8-13 用户管理数据访问层设计

表 8-10 UserDaoImpl 类

类名称	UserDaoImpl	中文名称	用户数据访问层类	类型	实体类	父类	dao.IUser
类描述	提供对用户管理数据访问层的 CRUD 各种功能操作，实现该层 IUser 接口，IUser 接口必须从 IDoData 继承						
方法							
序号	方法名	参数		返回类型			功能说明
1	add	AbsObject obj：新增的对象		int			将 obj 用户新增到数据库中，返回值大于 0 表示成功，其他表示失败
2	edit	AbsObject obj：修改的对象		int			对 obj 用户修改，返回值大于 0 表示成功，其他表示失败
3	del	AbsObject obj：删除的对象		int			将 obj 用户从数据库中删除，返回值大于 0 表示成功，其他表示失败
4	changeStatus	AbsObject obj：状态变更的对象		int			改变 obj 用户的状态，返回值大于 0 表示成功，其他表示失败
5	findResult	Map<String,Object> condition：查询条件		List<Map<String,Object>>			根据条件 condition 从数据库中查找数据，如果查询到数据，则返回 Map 结构数据，否则返回 null
6	find	Map<String,Object> condition：查询条件		List <AbsObject>			根据条件 condition 从数据库中查找数据，如果查询到数据，则返回 List 的用户数据，否则返回 null

续表

序号	方法名	方法 参数	返回类型	功能说明
7	getMe	String id：查询对象的 ID 编号	AbsObject	根据参数 ID，从数据库中查找，如果查找到，返回用户对象，否则返回 null

案例实现

根据案例分析，用户管理案例采用了三层架构与 MVC 设计理念，项目的后台代码目录结构如图 8-14 所示。

图 8-14 用户管理后台代码目录结构

1. 用户管理表示层

用户管理表示层采用了 MVC 设计理念，包含了视图层、控制层及模型层，各层的详细实现如下：

（1）视图层。

视图层设计了用户新增功能的 JSP 页面，其他功能的页面因篇幅所限未作实现。

① user.jsp 页面

user.jsp 页面提供了用户新增功能的操作页面，并实现了与控制层的通信，实现代码如下：

```jsp
<%--
  Created by IntelliJ IDEA.
  User: wph-pc
  Date: 2018/10/31
  Time: 22:53
  To change this template use File | Settings | File Templates.
--%>
<%@ page contentType="text/html;charset=UTF-8" language="java" %>
<%@include file="../header/init_bootstrap.jsp"%>
<html>
<head>
    <title>基于Servlet MVC用户信息采集</title>
```

```html
        </head>
        <body class="container">
        <div class="jumbotron">
            <h1>基于Servlet的MVC框架及三层架构用户管理</h1>
            <p>
                案例描述：利用Servlet技术及MVC理念，采用三层架构开发技术模拟用户基本信息的新增、修改、删除、查询等操作。
            </p>
        </div>
        <div class="row">
            <div class="col-md-12 col-sm-12 col-lg-12 col-xs-12">
                <form action="/jspweb/user" method="post">
                    <input type="hidden" name="doType" value="add">
                    <div class="form-group">
                        <label for="txtAccount">账号</label>
                        <input type="text" class="form-control" name="number" placeholder="请输入账号" id="txtAccount"/>
                    </div>
                    <div class="form-group">
                        <label for="txtName">真实姓名</label>
                        <input type="name" class="form-control" placeholder="请输入用户姓名" id="txtName"/>
                    </div>
                    <div class="form-group">
                        <label for="txtNickName">昵称</label>
                        <input type="nickName" class="form-control" placeholder="请输入用户昵称" id="txtNickName"/>
                    </div>
                    <button type="submit" class="btn btn-primary">新增</button>
                </form>
            </div>
        </div>
        </body>
        </html>
```

（2）模型层。

模型层实现了对用户数据模型的定义，是一个 JavaBean 类，根据设计，该类需要从 AbsObject 超级实体类继承。

① User 用户实体类

```java
        package business.entity;
        import business.entity.AbsObject;
        /**
         * 用户基本信息
         * Created by wph-pc on 2018/10/28.
         */
        public class User extends AbsObject {
            //用户账号
            private String number=null;
            //用户昵称
            private String nickName=null;
            //用户密码
```

```java
    private String password=null;

    public String getNumber() {
        return number;
    }

    public void setNumber(String number) {
        this.number = number;
    }

    public String getNickName() {
        return nickName;
    }

    public void setNickName(String nickName) {
        this.nickName = nickName;
    }

    public String getPassword() {
        return password;
    }

    public void setPassword(String password) {
        this.password = password;
    }
}
```

② AbsObject 超级实体类

```java
package business.entity;
import java.util.Date;

/**
 * 超级实体抽象类
 * Created by wph-pc on 2018/10/30.
 */
public abstract class AbsObject {
    //对象id号
    private String id=null;
    //对象名称
    private String name=null;
    //对象创建时间
    private Date createDate=null;
    //对象描述
    private String description=null;
    //对象状态
    private String status=null;
    public String getId() {
        return id;
    }
    public void setId(String id) {
        this.id = id;
    }
```

```java
    public String getName() {
        return name;
    }
    public void setName(String name) {
        this.name = name;
    }
    public Date getCreateDate() {
        return createDate;
    }
    public void setCreateDate(Date createDate) {
        this.createDate = createDate;
    }
    public String getDescription() {
        return description;
    }
    public void setDescription(String description) {
        this.description = description;
    }
    public String getStatus() {
        return status;
    }
    public void setStatus(String status) {
        this.status = status;
    }
}
```

③ KesunReturn 返回类

```java
package business.entity;

/**
 * 操作返回实体类
 * Created by wph-pc on 2018/10/30.
 */
public class KesunReturn {
    //返回状态码
    private String code=null;
    //返回提示信息
    private String message=null;
    //返回操作结果
    private Object obj=null;

    public String getCode() {
        return code;
    }

    public void setCode(String code) {
        this.code = code;
    }

    public String getMessage() {
        return message;
    }
}
```

```java
    public void setMessage(String message) {
        this.message = message;
    }

    public Object getObj() {
        return obj;
    }

    public void setObj(Object obj) {
        this.obj = obj;
    }
}
```

（3）控制层。

控制层由 KesunSuperController 控制层超级类、IUser 接口和 UserController 用户控制层子类组成。

① KesunSuperController 控制层超级类

```java
package business.controller;
import business.entity.AbsObject;
import business.entity.KesunReturn;
import business.service.KesunSuperService;

import javax.Servlet.http.HttpServlet;
import javax.Servlet.http.HttpServletRequest;
import java.io.IOException;
import java.util.Map;
import java.util.Properties;

/**
 * 控制层超级类
 * Created by wph-pc on 2018/10/30.
 */
public abstract class KesunSuperController extends HttpServlet {
    //定义业务层操作对象变量
    private KesunSuperService bll=null;
    public KesunSuperService getBll() {
        return bll;
    }
    public void setBll(KesunSuperService bll) {
        this.bll = bll;
    }

    //根据参数创建实体对象
    public abstract AbsObject createModel(HttpServletRequest request);
    //根据参数创建查询条件
    public abstract Map<String,Object> createCondition(HttpServletRequest request);
    /*判断业务对象bll是否有效*/
    private KesunReturn judgeService()
    {
```

```java
        KesunReturn back=new KesunReturn();
        if (bll==null){
            back.setCode("0");
            back.setMessage("业务对象为null");
            back.setObj(null);
        }
        else
            back.setObj(bll);
        return back;
    }
    public KesunReturn add(HttpServletRequest request){
        KesunReturn back=judgeService();
        if (back.getObj()==null) return back;
        bll.setModel(createModel(request));
        return bll.add();
    }

    public KesunReturn edit(HttpServletRequest request){
        KesunReturn back=judgeService();
        if (back.getObj()==null) return back;
        bll.setModel(createModel(request));
        return bll.edit();
    }
    public KesunReturn del(HttpServletRequest request){
        KesunReturn back=judgeService();
        if (back.getObj()==null) return back;
        bll.setModel(createModel(request));
        return bll.del();
    }

    public KesunReturn changeStatus(HttpServletRequest request){
        KesunReturn back=judgeService();
        if (back.getObj()==null) return back;
        bll.setModel(createModel(request));
        return bll.changeStatus();
    }
    public KesunReturn getMe(HttpServletRequest request){
        KesunReturn back=judgeService();
        if (back.getObj()==null) return back;
        bll.setModel(createModel(request));
        return bll.getMe();
    }

    public KesunReturn find(HttpServletRequest request){
        KesunReturn back=judgeService();
        if (back.getObj()==null) return back;
        Map<String,Object> cons=createCondition(request);
        return bll.find(cons);
    }
    public KesunReturn findForMap(HttpServletRequest request){
```

```java
            KesunReturn back=judgeService();
            if (back.getObj()==null) return back;
            Map<String,Object> cons=createCondition(request);
            return bll.findForMap(cons);
    }
        /*根据业务层类名称创建业务对象*/
        private KesunSuperService createService(String moduleName)
        {
            Properties bprops=new Properties();
            try {
                bprops.load(KesunSuperController.class.getClassLoader().getResourceAsStream("business.properties"));
            } catch (IOException e) {
                e.printStackTrace();
                return null;
            }
            String serviceClassName=bprops.get(moduleName).toString();
            KesunSuperService service=null;
            try {
                Class bll=Class.forName(serviceClassName);
                try {
                    service=(KesunSuperService)bll.newInstance();
                } catch (InstantiationException e) {
                    e.printStackTrace();
                    return null;
                } catch (IllegalAccessException e) {
                    e.printStackTrace();
                    return null;
                }
            } catch (ClassNotFoundException e) {
                e.printStackTrace();
                return null;
            }
            return service;
        }

}
```

② controller.IUser 接口

```java
package business.controller;

/**
 * 控制层用户接口
 * Created by wph-pc on 2018/11/1.
 */
public interface IUser {
}
```

③ UserController 类

```java
package business.controller.impl;

import business.controller.IUser;
import business.controller.KesunSuperController;
```

```java
import business.entity.KesunReturn;
import business.entity.User;
import business.entity.AbsObject;
import business.service.impl.UserServiceImpl;

import javax.Servlet.http.HttpServletRequest;
import javax.Servlet.http.HttpServletResponse;
import java.io.IOException;
import java.util.Map;

/**
 * 用户控制层类
 * Created by wph-pc on 2018/10/31.
 */
public class UserController extends KesunSuperController implements IUser {
    public UserController()
    {
        //实例化用户业务层对象
        UserServiceImpl bll=new UserServiceImpl();
        //设置控制层的用户业务对象
        setBll(bll);
    }
    /*根据用户页面请求参数生成用户对象*/
    @Override
    public AbsObject createModel(HttpServletRequest request) {
        //获取控制层用户实体对象
        User u=(User)getBll().getModel();
        /*设置用户实体各种属性信息*/
        u.setNumber(request.getParameter("number"));
        u.setNickName(request.getParameter("nickName"));
        u.setName(request.getParameter("name"));
        return u;
    }

    /*根据用户页面查询参数,转换成Map结构参数*/
    @Override
    public Map<String, Object> createCondition(HttpServletRequest request) {
        return null;
    }
    @Override
    public void doPost(HttpServletRequest request, HttpServletResponse response) throws IOException {
        KesunReturn back=new KesunReturn();
        back.setMessage("系统没有获取到任何操作!");
        back.setCode("0");
        back.setObj(null);
        /*设置编码格式及响应类型*/
        request.setCharacterEncoding("utf-8");
        response.setContentType("text/html");
```

```java
            response.setCharacterEncoding("utf-8");

            //获取客户端请求的操作类型
            String doType=request.getParameter("doType");
            switch (doType){
                case "add":
                    back=getBll().add();
                    break;
                default:
                    back.setMessage("抱歉,系统没有获取到任何的操作指令!");
                    break;
            }
            response.getWriter().write(back.getMessage());
        }
    }
```

控制层用户 UserController 类实现后,需要在 web.xml 文件中配置 Servlet 节点,配置信息如下:

```xml
    <Servlet>
        <Servlet-name>user</Servlet-name>
        <Servlet-class>business.controller.impl.UserController</Servlet-class>
    </Servlet>
    <Servlet-mapping>
        <Servlet-name>user</Servlet-name>
        <url-pattern>/user</url-pattern>
    </Servlet-mapping>
```

2. 用户管理业务层实现

用户管理业务层实现由该层 KesunSuperService 业务层超级类、service.IUser 接口及 UserServiceImpl 类组成。

① KesunSuperService 业务层超级类

```java
    package business.service;

    import business.entity.AbsObject;
    import business.entity.KesunReturn;
    import business.dao.IDoData;

    import java.util.List;
    import java.util.Map;

    /**
     * 业务层超级类
     * Created by wph-pc on 2018/10/30.
     */
    public abstract class KesunSuperService {
        public AbsObject getModel() {
            return model;
        }
        public void setModel(AbsObject model) {
            this.model = model;
```

```java
}
    private AbsObject model=null;
    
    /*获取数据访问层对象*/
    public abstract IDoData getDAO();
    
    /*数据访问层和实体对象条件检测*/
    private KesunReturn checkCondition(){
        KesunReturn back=new KesunReturn();
        if (getDAO()==null || model==null){
            back.setCode("0");
            back.setMessage("数据访问层对象或数据模型都不能为空!");
            back.setObj(null);
        }else{
            back.setCode("1");
            back.setMessage("数据访问层对象和数据模型满足操作要求!");
            back.setObj(1);
        }
        return back;
    }
    /*对象新增*/
    public KesunReturn add() {
        KesunReturn back=checkCondition();
        if (back.getObj()==null) return back;
        try {
            int result=getDAO().add(model);
            if (result<=0)
            {
                back.setCode("0");
                back.setMessage("抱歉，您的数据保存失败!");
            }
            else
            {
                back.setCode("1");
                back.setMessage("您的数据已经成功提交!");
            }
            back.setObj(result);
        } catch (Exception e) {
            e.printStackTrace();
            back.setCode("-1");
            back.setMessage("数据保存出现异常,异常信息: "+e.getMessage());
            back.setObj(-1);
        }
        return back;
    }
    /*对象修改*/
    public KesunReturn edit() {
        KesunReturn back=checkCondition();
        if (back.getObj()==null) return back;
        int result = 0;
        try {
```

```java
            result = getDAO().edit(model);
            if (result<=0)
            {
                back.setCode("0");
                back.setMessage("数据修改操作失败！");
            }
            else
            {
                back.setCode("1");
                back.setMessage("数据修改成功！");
            }
            back.setObj(result);
        } catch (Exception e) {
            e.printStackTrace();
            back.setObj(-1);
            back.setMessage("修改失败，出现异常信息："+e.getMessage());
            back.setCode("-1");
        }
        return back;
    }
    /*对象删除*/
    public KesunReturn del()
    {
        KesunReturn back=new KesunReturn();
        //检测操作条件
        if (back.getObj()==null) return back;
        
        int result=0;
        try {
            result=getDAO().del(model);
            if (result>0)
                back.setMessage("您指定的数据已经成功删除！");
            else
                back.setMessage("删除失败，系统返回错误码："+result);
        } catch (Exception e) {
            e.printStackTrace();
            back.setMessage("系统出现异常信息："+e.getMessage());
            result=-1;
        }
        back.setCode(String.valueOf(result));
        back.setObj(result);
        return back;
    }
    
    /*检测查询的条件*/
    private KesunReturn checkFindCondition(Map condition){
        KesunReturn back=new KesunReturn();
        if (getDAO()==null || condition==null){
            back.setCode("0");
            back.setMessage("数据访问层对象或查找都不能为空！");
            back.setObj(null);
```

```java
        }else{
            back.setCode("1");
            back.setMessage("数据访问层对象和查找条件满足操作要求！");
            back.setObj(1);
        }
        return back;
    }
    /*判断查询条件param是否有效，返回KesunReturn结果*/
    private KesunReturn getFindResult(List param){
        KesunReturn back=new KesunReturn();
        if (param==null || param.size()==0)
        {
            back.setCode("0");
            back.setMessage("抱歉，系统没有查询到符合条件的数据！");
        }
        else
        {
            back.setCode("1");
            back.setMessage("系统已经查询到符合条件的数据！");
        }
        back.setObj(param);
        return back;
    }
    /**
     * 根据条件查找对象
     * @return
     * @throws Exception
     */
    public KesunReturn find(Map values){
        KesunReturn back=checkFindCondition(values);
        if (back.getObj()==null) return back;
        try {
            List<AbsObject> temp=getDAO().find(values);
            //转换查询结果
            back=getFindResult(temp);
        } catch (Exception e) {
            e.printStackTrace();
            back.setCode("-1");
            back.setMessage("系统查询的过程中出现异常信息："+e.getMessage());
            back.setObj(null);
        }
        return back;
    }
    /*对象查找*/
    public KesunReturn findForMap(Map values) {
        KesunReturn back=new KesunReturn();
        try {
            List<Map<String,Object>> temp=getDAO().findResult(values);
            //转换查询结果
            back=getFindResult(temp);
        } catch (Exception e) {
```

```java
            back.setCode("-1");
            back.setMessage("系统处理查询的过程中出现异常,异常信息: "+e.getMessage());
            back.setObj(null);
            e.printStackTrace();
        }
        return back;
    }
    /*单个对象查找*/
    public KesunReturn getMe(){
        //检测数据访问层对象DAO及model有效性
        KesunReturn back=checkCondition();
        if (back.getObj()==null) return back;
        try {
            AbsObject temp=getDAO().getMe(model.getId());
            if (temp==null)
            {
                back.setCode("0");
                back.setMessage("系统没有查询到您需要的数据!");
            }
            else
            {
                back.setCode("1");
                back.setMessage("系统已经查询到您需要的数据,在obj对象中!");
            }
            back.setObj(temp);
        } catch (Exception e) {
            back.setCode("-1");
            back.setMessage("系统异常: "+e.getMessage());
            back.setObj(null);
        }
        return back;
    }
    /*对象状态变更*/
    public KesunReturn changeStatus() {
        //检测条件DAO和model的有效性
        KesunReturn result = checkCondition();
        if (result.getObj()==null) return result;
        try {
            int temp=getDAO().changeStatus(model);
            result.setCode(String.valueOf(temp));
            if (temp>0)
                result.setMessage("状态变更成功! ");
            else
                result.setMessage("状态变更失败! ");
            result.setObj(temp);

        } catch (Exception e) {
            result.setCode("-1");
            result.setMessage("系统出现异常: "+e.getMessage());
            result.setObj(null);
```

```
            }
            return result;
        }
    }
```

② service.IUser 接口

```
package business.service;

/**
 * 业务层用户接口
 * Created by wph-pc on 2018/11/1.
 */
public interface IUser {
}
```

③ UserServiceImpl 类

```
package business.service.impl;
import business.dao.IDoData;
import business.dao.impl.UserDaoImpl;
import business.entity.User;
import business.service.IUser;
import business.service.KesunSuperService;
/**
 * 用户管理业务层类
 * Created by wph-pc on 2018/10/31.
 */
public class UserServiceImpl extends KesunSuperService implements IUser {
    public UserServiceImpl()
    {
        //设置持久层User对象
        setModel(new User());
    }
    @Override
    public IDoData getDAO() {
        //创建User数据访问对象
        return new UserDaoImpl();
    }
}
```

3. 用户管理数据访问层实现

用户管理数据访问实现由 IDoData 数据处理接口、dao.IUser 用户业务接口及 UserDaoImpl 用户数据访问类组成，其中 UserDaoImpl 类中的功能是模拟，如果读者学习了数据库，可以将有关数据库操作在对象的方法中替换。

① IDoData 数据处理接口

```
package business.dao;
import business.entity.AbsObject;
import java.util.List;
import java.util.Map;
/**
 * 数据访问层操作接口
 * Created by wph-pc on 2018/10/30.
```

```java
 */
public interface IDoData {
    //对象新增
    int add(AbsObject obj);
    //对象修改
    int edit(AbsObject obj);
    //对象删除
    int del(AbsObject obj);
    //对象状态变更
    int changeStatus(AbsObject obj);
    //查找对象,以二维表格返回
    List<Map<String,Object>> findResult(Map<String,Object> condition);
    //查找对象,以对象集合返回
    List<AbsObject> find(Map<String,Object> condition);
    //根据ID查找对象
    AbsObject getMe(String id);
}
```

② dao.IUser 用户业务接口

```java
package business.dao;

/**
 * 数据访问层用户接口
 * Created by wph-pc on 2018/11/1.
 */
public interface IUser extends IDoData {
}
```

③ UserDaoImpl 用户数据访问类

```java
package business.dao.impl;
import business.dao.IDoData;
import business.dao.IUser;
import business.entity.User;
import business.entity.AbsObject;
import java.util.ArrayList;
import java.util.HashMap;
import java.util.List;
import java.util.Map;

/**
 * 用户操作数据访问层模拟类
 * Created by wph-pc on 2018/10/31.
 */
public class UserDaoImpl implements IUser {
    @Override
    public int add(AbsObject obj) {
        //下面是模拟数据,如果您学习了数据库编程,替换成数据库操作
        return 1;
    }

    @Override
    public int edit(AbsObject obj) {
        //下面是模拟数据,如果您学习了数据库编程,替换成数据库操作
```

```java
        return 1;
    }

    @Override
    public int del(AbsObject obj) {
        //下面是模拟数据，如果您学习了数据库编程，替换成数据库操作
        return 1;
    }

    @Override
    public int changeStatus(AbsObject obj) {
        //下面是模拟数据，如果您学习了数据库编程，替换成数据库操作
        return 1;
    }

    @Override
    public List<Map<String, Object>> findResult(Map<String, Object> condition) {
        //下面是模拟数据，如果您学习了数据库编程，替换成数据库查找
        List<Map<String,Object>> result=new ArrayList<>();
        /*设置账号信息*/
        Map<String,Object> number=new HashMap<>();
        number.put("number","20181101");
        result.add(number);

        /*设置昵称信息*/
        Map<String,Object> nickName=new HashMap<>();
        number.put("nickName","kingboy");
        result.add(nickName);

        /*设置姓名*/
        Map<String,Object> name=new HashMap<>();
        number.put("name","王平华");
        result.add(name);
        return result;
    }

    @Override
    public List<AbsObject> find(Map<String, Object> condition) {
        //下面是模拟数据，如果您学习了数据库编程，替换成数据库查找
        List<AbsObject> result=new ArrayList<>();
        /*设置账号信息*/
        User user=new User();
        user.setNumber("20181101");
        user.setName("王平华");
        user.setNickName("kingboy");
        result.add(user);
        return result;
    }

    @Override
```

```
public AbsObject getMe(String id) {
    //下面是模拟数据，如果您学习了数据库编程，替换成数据库操作
    /*设置账号信息*/
    User user=new User();
    user.setNumber("20181101");
    user.setName("王平华");
    user.setNickName("kingboy");
    return user;
}
```

运行结果

按照本节案例的要求实现所有代码后，启动视图层中的 user.jsp 页面即可以看到运行结果。

单击图 8-15 中的"新增"按钮，显示如图 8-16 所示的结果。

图 8-15　用户管理运行页面

图 8-16　用户新增运行结果页面

本章通过用户身份验证案例讲解了 Servlet 技术的基本使用规则，而项目案例用户管理模块充分发挥了 Servlet 在实际项目设计中的使用。通过 Servlet 在 MVC 中控制层扮演的角色，读者更能理解 Servlet 的作用，为后续的 J2EE 开发打下了坚实的基础。另外，三层架构是目前项目设计的主要框架，通过用户管理模块进行了翔实的讲解，并配以丰富的图解和表格描述，加深了三层架构理念的理解。

习题

安全管理在系统开发中非常重要，通过功能授权给操作者是安全操作的一种非常重要

的方式，操作者在使用系统功能时，系统判断当前操作者是否拥有该功能，如果有就执行该功能，否则就取消该功能的执行。在实行功能安全管理前，需要对功能权限的基本信息进行管理，包括功能权限的新增、修改、删除及查询操作等。具体操作要求如下：

1．功能权限基本信息包括的属性有：编号、名称、权限地址、上级权限等；采用 JavaBean 类进行实体类定义。

2．功能权限的基本信息维护要求采用三层架构技术，定义表示层、业务层及数据访问层；其中业务层及数据访问层需要定义接口。

3．功能权限的业务层类必须从 KesunSuperService 业务层超级类继承，功能权限数据访问层接口必须从 IDoData 继承。

4．功能权限基本信息维护的表示层要求采用 MVC 机制实现。

5．提供功能权限基本信息新增、修改、删除及查询操作的页面。

6．功能权限数据访问层中的各功能方法可以为空或暂时模拟实现。

第 9 章

过滤器与监听器

在开发 Java Web 项目过程中，经常需要各种客户端对服务器端的请求进行验证，例如：需要登录后才能访问的页面，对一些没有访问权限的页面取消其访问功能等；JSP 中提供了过滤器与监听器实现这些安全及权限验证的功能。过滤器可以过滤一些不安全或非法访问请求；而监听器主要负责监听来自客户端的各种访问请求，包括会话监听、数据请求监听、会话属性变更监听等。本章主要讲解过滤器及监听的相关知识，通过《日志采集》项目案例及《单点登录及权限应用》案例，阐述它们在项目中的应用。

本章任务

（1）过滤器及监听器的相关概念；
（2）日志采集；
（3）单点登录及权限应用。

重点内容

（1）掌握过滤器及监听器接口中的事件；
（2）掌握过滤器及监听器在项目中的应用。

难点内容

（1）过滤器在项目中应用；
（2）监听器在项目中应用

9.1 过滤器与监听器相关知识

过滤器与监听器在 Java Web 项目中非常重要，在系统安全及权限管理方面起到无可替代的作用，在应用过滤器及监听器之前，需要对过滤器及监听器的相关基础知识有所了解。

9.1.1 过滤器

1．过滤器相关知识

JSP 中过滤器是一种对客户端请求进行过滤处理的机制，它以接口 Filter 形式存在，使用时需要实现该接口。过滤器应用广泛，例如：防止用户未登录访问需要身份验证的请求，可以通过设置过滤器的形式进行处理。过滤器在身份验证中的作用如图 9-1 所示。

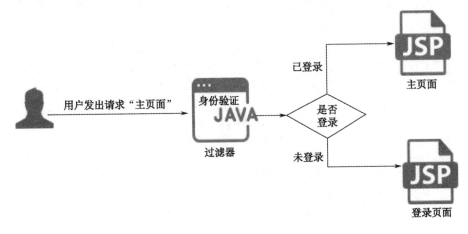

图 9-1 过滤器在身份验证中的作用

在图 9-1 中，用户向服务器页面发出请求访问主页面，请求必须先通过"身份验证过滤器"，该过滤器验证当前用户是否经过合法的身份验证，如果已经验证过，则可以直接访问主页面；如果没有经过身份验证，则过滤器将请求转给登录页面，登录页面成功后，身份验证过滤器再验证，合法后则跳转到目标主页面。

过滤器理论上可以拦截一切客户端请求，当然也可以设置拦截部分请求。JSP 中定义了过滤器的数据处理接口 Filter，各种过滤器实体类必须实现该接口，Filter 接口描述见表 9-1。

表 9-1 Filter 接口描述

类 名 称	Filter	中文名称		过滤器	类 型	接口
类 描 述	提供实体过滤器类的各种方法功能定义					
序 号	方 法 名	参 数		返回类型	功能说明	
1	init	FilterConfig var1：过滤参数		void	过滤器初始化，通常可以配置过滤器的参数	
2	doFilter	ServletRequest var1：请求对象；ServletResponse var2：服务器响应对象；FilterChain var3：过滤链对象；		void	过滤器的核心功能，所有过滤器拦截的请求都会经过该方法，负责处理拦截对象的处理功能	
3	destroy	无		void	过滤器销毁执行的方法	

在表 9-1 中，方法 init 是在过滤器实例化之前执行的方法，实现过滤器中配置参数的读取，参数 FilterConfig 是一个接口，其接口描述见表 9-2。

表 9-2 FilterConfig 接口描述

类名称	FilterConfilg	中文名称	过滤器配置	类型	接口
类描述	提供过滤器参数读取操作的定义				
序号	方法名	参数	返回类型	功能说明	
1	getFilterName	无	string	获取过滤器名称	
2	getServletContext	无	ServletContext	获取上下文对象	
3	getInitParameter	String var1	string	根据参数名称，获取过滤器初始化参数值	
4	getInitParameterNames	无	Enumeration<String>	获取过滤器所有初始化参数名称	

例如，读取页面编码格式的过滤器参数，如图 9-2 所示。

```
<filter>
    <filter-name>encoding</filter-name>
    <filter-class>chapter9.EncodingFilter</filter-class>
    <init-param>
        <param-name>encoding</param-name>
        <param-value>utf-8</param-value>
    </init-param>
</filter>
```

图 9-2 编码格式过滤器配置

图 9-2 中，过滤器名称 encoding，处理过滤器类 chapter9.EncodingFilter，并配置初始化参数，参数名称为 encoding，初始化值为 utf-8。通过 FilterConfig 接口中的 getFilterName 方法获取到名称 encoding；通过 FilterConfig 接口中的 getInitParameter 方法，给定参数名 "encoding"，可以获取到 "utf-8" 值。

2. 过滤器开发步骤

过滤器的开发步骤如下：

第一步，开发过滤器实体类，实现接口 Filter；格式如下：

```java
public class 类名称 implements Filter {
@Override
    public void init(FilterConfig filterConfig) throws ServletException {
        //过滤器初始化方法
    }
    @Override
    public void doFilter(ServletRequest ServletRequest, ServletResponse ServletResponse, FilterChain filterChain) throws IOException, ServletException {
        //过滤方法
    }
    @Override
    public void destroy() {
        //过滤器销毁方法
    }
}
```

第二步，向应用程序注册过滤器。在 web.xml 中，配置过滤器节点：

```
<filter>
    <filter-name>
```

```
                自定义过滤器对象名
        </filter-name>
    <filter-class>
                定义过滤器类,含包名
        </filter-class>
</filter>
```

第三步,设置过滤器的过滤路径。向应用程序注册过滤器后,还需要配置过滤路径,设置哪些请求需要过滤,过滤器过滤路径配置在 web.xml 中,需要完成第二步才能进行设置:

```
<filter-mapping>
    <filter-name>
            自定义过滤器对象名称,与 filter 注册名称一致
        </filter-name>
    <url-pattern>
            过滤的路径,可以是页面路径,也可以是 Servlet
        </url-pattern>
</filter-mapping>
```

特别说明:如果同一个过滤器需要过滤多个路径,可以配置多个 filter-mapping 节点。

9.1.2 监听器

1. 相关知识

监听器是 Servlet 规范中定义的一种特殊类,可以用来监听 ServletContext、HttpSession 和 ServletRequest 等域对象的创建和销毁事件;还可以监听这些域对象的增加或删除事件,可以在事件发生前或发生后做一些必要的处理。

监听器在项目中的作用不可忽略,例如利用监听器可以进行在线人数统计、用户的单点登录控制等。

监听器按照监听事件可以分为:
- 监听会话对象创建与销毁事件的会话监听器 HttpSessionListener;
- 监听会话对象属性创建与销毁事件的会话属性监听器 HttpSessionAttributeListener;
- 监听上下文对象创建与销毁事件的上下文监听器 ServletContextListener;
- 监听客户端请求对象创建与销毁事件的请求监听器 ServletRequestListener;

上面四类监听器都是从 EventListener 接口中继承过来的,监听器与父类之间的关系如图 9-3 所示。

图 9-3 监听器与父类之间的关系

图 9-3 中,接口 EventListener 是一个无任何方法的空接口,作为四个子接口的父类。

会话监听器 HttpSessionListener 接口从 EventListener 继承,当客户端与服务器建立会话时,该监听器就会知道,利用这个机制可以实现用户在线统计的功能。HttpSessionListener 接口描述见表 9-3。

表 9-3　HttpSessionListener 接口描述

类 名 称	HttpSessionListener	中文名称	会话监听器	类　型	接口，继承 EventListener
类 描 述	提供对 session 会话对象的创建及销毁的监听定义					
序　号	方 法 名	参　数	返回类型	功能说明	
1	sessionCreated	HttpSessionEvent var	void	session 会话创建时执行的方法	
2	sessionDestroyed	HttpSessionEvent var	void	session 销毁时执行的方法	

会话属性监听器 HttpSessionAttributeListener 接口从 EventListener 继承过来，用来监听 Session 对象的属性新增、删除及属性变更，HttpSessionAttributeListener 接口描述见表 9-4。

表 9-4　HttpSessionAttributeListener 接口描述

类 名 称	HttpSessionAttributeListener	中文名称	会话属性监听器	类　型	接口，继承 EventListener
类 描 述	提供对 session 会话对象的属性新增、删除及值变更的监听定义					
序　号	方 法 名	参　数	返回类型	功能说明	
1	attributeAdded	HttpSessionBindingEvent var	void	属性新增时执行的方法	
2	attributeRemoved	HttpSessionBindingEvent var	void	属性删除时执行的方法	
3	attributeReplaced	HttpSessionBindingEvent var	void	属性值变更时执行的方法	

上下文监听器 ServletContextListener 接口从 EventListener 继承，用来监听上下文初始化及销毁，通过该监听器，可以获取 web.xml 文件中的上下文对象的各种配置信息。ServletContextListener 接口描述见表 9-5。

表 9-5　ServletContextListener 接口描述

类 名 称	ServletContextListener	中文名称	上下文监听器	类　型	接口，继承 EventListener
类 描 述	提供对上下文对象的初始化及销毁的监听定义					
序　号	方 法 名	参　数	返回类型	功能说明	
1	contextInitialized	ServletContextEvent var	void	上下文对象初始化时执行的方法	
2	contextDestroyed	ServletContextEvent var	void	上下文对象销毁时执行的方法	

请求监听器 ServletRequestListener 接口从 EventListener 继承，实现对 Servlet 请求进行监听，对请求的初始化和销毁进行监听。ServletRequestListener 接口描述见表 9-6。

表 9-6　ServletRequestListener 接口描述

类 名 称	ServletRequestListener	中文名称	请求监听器	类　型	接口，继承 EventListener
类 描 述	提供对请求对象的初始化及销毁的监听定义					
序　号	方 法 名	参　数	返回类型	功能说明	
1	requestInitialized	ServletRequestEvent var	void	请求对象初始化时执行的方法	
2	requestDestroyed	ServletRequestEvent var	void	请求对象销毁时执行的方法	

2. 监听器开发步骤

监听器的开发步骤如下：

- 监听器类开发，需要实现 HttpSessionListener 会话监听器、HttpSessionAttributeListener 会话属性监听器、ServletContextListener 上下文监听器和 ServletRequestListener 请求监听器等四个监听器接口中的任意一个。定义语法格式：

```
public class 监听器类名称 implements 监听器接口 {
    //实现接口中的各个方法
}
```

- 向应用程序注册监听器，在 web.xml 中配置监听器节点，一个应用程序可以配置多个监听器。监听器配置完成后，会对有所的对应程序进行监听，其语法格式如下：

```
<listener>
    <listener-class>监听器类，含完整的包名称</listener-class>
</listener>
```

9.2 日志文件

日志文件一般有两种，一种是用来监控程序运行的状况；另外一种是用来记录系统用户访问应用程序的情况。本章的日志文件案例是第二种，通过采集用户操作应用程序，分析应用程序各种功能模块的使用情况、用户习性，为将来大数据的采集应用打下基础。

案例描述

利用 JSP 的过滤器机制，开发能够详细采集用户访问应用程序的情况，具体的日志内容包括：用户访问设备类型、操作系统、访问时间、访问模块和 IP 地址等，并将这些采集的数据写入文件中进行存储。具体要求：

- 采集所有用户访问服务器的记录信息；
- 记录信息至少需要包括：客户端的 IP 地址、操作系统、浏览器版本、访问模块功能名称、访问时间、会话 ID 及操作用户等；
- 将采集到的日志信息保存到服务器端指定路径下的文件中。

案例分析

根据案例描述，需要解决三个问题：用户访问服务器所有的记录、确定日志信息的结构及将采集到的信息写入服务器指定路径下的文件中。日志文件解决思路如图 9-4 所示。

图 9-4 中，用户通过浏览器访问服务端的任何 Web 应用程序，必然会通过 web.xml 配置，设计思路：

- 设计一个日志过滤器，负责拦截所有的用户请求；并形成日志信息；
- 将所有规定的日志信息设计成一个数据实体模型；
- 当拦截到用户请求后，形成日志信息对象，通过 log4j 框架将日志信息写入服务器指定文件中。

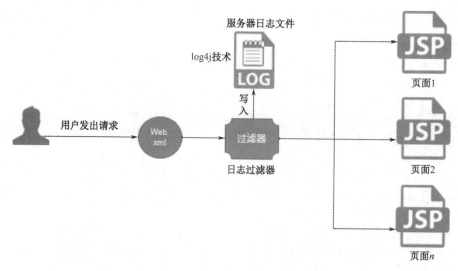

图 9-4　日志文件解决思路

按照以上思路，用户访问提请任何的请求都会被日志过滤器拦截，拦截到后，先将用户请求形成日志对象，写入服务器日志文件中，再将用户请求转向指定目标。

1. 日志过滤器

日志过滤器是日志文件中的核心，负责拦截用户请求。日志过滤器设计成一个类 LogFilter，其设计见表 9-7。

表 9-7　LogFilter 类设计

类名称	LogFilter	中文名称	日志过滤器类	类　型	实体类	父类	Filter
类描述	实现采集用户访问服务器记录信息						
成员变量							
序　号	变量名	修饰词	类　型		作　用		
1	logger	static	org.apache.log4j.Logger		私有变量，提供 log4j 对象		
方　法							
序　号	方法名	参　数		返回类型	功能说明		
1	getCustomerLog	ServletRequest request：请求对象		string	将请求对象参数 request 转成日志字符串		
2	init	FilterConfig filterConfig：过滤配置参数对象		void	继承实现父接口中的方法，过滤器初始化后，可以获取过滤器参数配置信息。此处应用为空		
3	doFilter	ServletRequest ServletRequest：页面请求对象；ServletResponse ServletResponse：页面响应对象；FilterChain filterChain：过滤链对象		void	继承实现父接口中的方法，核心内容，里面实现日志文件的数据写入		
4	destroy	无		void	继承实现父接口中的方法，此处的应用为空		

2. 日志信息实体模型类

日志信息实体模型类将需要保存在日志文件中的日志信息集中管理，形成日志实体模型，LogFilter 类设计见表 9-8。

表 9-8　LogFilter 类设计

类名称	Log	中文名称	日志信息实体模型类	类型	实体类	父类	无
类描述	实现对日志信息的表示						
成员变量							
序号	变量名	访问权限		类型		作用	
1	os	private		string		记录操作系统信息	
2	ipAddress port	private		string		记录客户端的 IP 地址	
3	port	private		string		记录客户端的访问端口	
4	browser	private		string		记录客户端浏览器的信息	
5	browserInfo	private		string		记录客户端浏览器的版本信息	
6	protocal	private		string		记录客户端访问协议	
7	sessionId	private		string		记录客户端 session 的 ID	
8	url	private		string		记录客户端访问模块地址	
9	createDate	private		date		记录访问时间	
10	user	private		string		记录访问者	

案例实现

日志文件实现，按照以下步骤操作：

1. 定义 Log 类

```java
package chapter9;
import java.util.Date;
/**
 * 客户访问请求日志数据模型
 * Created by wph-pc on 2018/11/6.
 */
public class Log {
    //操作系统
    private String os;
    //IP地址
    private String ipAddress;
    //端口号
    private String port;
    //浏览器
    private String browser;
    //浏览器版本号
    private String browserInfo;
    //访问协议
    private String protocal;
    //session
```

```java
    private String sessionId;
    //请求地址
    private String url;
    //请求时间
    private Date createDate=new Date();
    //请求用户
    private String user;
    public String getOs() {
        return os;
    }
    public void setOs(String os) {
        this.os = os;
    }
    public String getIpAddress() {
        return ipAddress;
    }
    public void setIpAddress(String ipAddress) {
        this.ipAddress = ipAddress;
    }
    public String getPort() {
        return port;
    }
    public void setPort(String port) {
        this.port = port;
    }
    public String getBrowser() {
        return browser;
    }
    public void setBrowser(String browser) {
        this.browser = browser;
    }
    public String getBrowserInfo() {
        return browserInfo;
    }
    public void setBrowserInfo(String browserInfo) {
        this.browserInfo = browserInfo;
    }
    public String getProtocal() {
        return protocal;
    }
    public void setProtocal(String protocal) {
        this.protocal = protocal;
    }
    public String getSessionId() {
        return sessionId;
    }
    public void setSessionId(String sessionId) {
        this.sessionId = sessionId;
    }
    public String getUrl() {
        return url;
```

```
    }
    public void setUrl(String url) {
        this.url = url;
    }
    public Date getCreateDate() {
        return createDate;
    }
    public void setCreateDate(Date createDate) {
        this.createDate = createDate;
    }
    public String getUser() {
        return user;
    }
    public void setUser(String user) {
        this.user = user;
    }
    @Override
    public String toString(){
        return "操作系统："+os+" IP地址："+ipAddress+" 端口号："+port+" 访问协议："+protocal+
                " 浏览器："+browser+" 浏览器版本号："+browserInfo+" sessionID："+getSessionId()+
                " 请求模块："+url.substring(url.lastIndexOf("/"))+" 请求用户："+user+
                " 访问时间："+getCreateDate();
    }
}
```

Log 日志实体类是一个 JavaBean 类，每个成员变量提供了 get 和 set 方法进行数据的读写操作。

2. 配置 log4j

Log4j 是 Apache 组织机构的开源日志框架，通过该框架将获取的日志信息写入日志文件中。

- 从 Apache 官网上下载 log4j 的 jar 包，并配置到项目 lib 库中；
- 配置 log4j 的基本配置信息，通过配置属性文件实现，在项目的资源文件夹中添加属性文件 log4j.properties；
- 配置 log4j.properties 属性文件内容信息：

```
### 设置Log4j输出配置###
log4j.rootLogger = INFO,C,D
### 控制台输出配置 ###
log4j.appender.C = org.apache.log4j.ConsoleAppender
log4j.appender.C.Target = System.out
log4j.appender.C.layout = org.apache.log4j.PatternLayout
log4j.appender.C.layout.ConversionPattern = [%p] [%d{yyyy-MM-dd HH:mm:ss}] %C.%M(%L) | %m%n
### 日志文件保存设置 ###
log4j.appender.D = org.apache.log4j.DailyRollingFileAppender
log4j.appender.D.File =d://jsplog/info.log
log4j.appender.D.Append = true
```

```
log4j.appender.D.Threshold = INFO
log4j.appender.D.layout = org.apache.log4j.PatternLayout
log4j.appender.D.layout.ConversionPattern = [%p] [%-d{yyyy-MM-dd HH:mm:ss}] %C.%M(%L) | %m%n
```

在配置文件中，配置了文件的写入地址"d://jsplog/info.log"，此外还配置了控制台输出信息。

3. 过滤器 LogFilter 类

过滤器 LogFilter 类是日志文件的核心，实现了用户请求的拦截及将访问日志信息写入服务器指定路径下的文件中。LogFilter 类的定义：

```java
package chapter9;
import business.entity.User;
import chapter6.requestout.TerminateDevice;
import org.apache.log4j.*;
import javax.Servlet.*;
import javax.Servlet.Filter;
import javax.Servlet.http.HttpServletRequest;
import java.io.Console;
import java.io.IOException;
/**
 * 客户端日志收集过滤器
 * Created by wph-pc on 2018/11/6.
 */
public class LogFilter implements Filter {
    //创建日志logger对象
    private static org.apache.log4j.Logger logger=Logger.getLogger(LogFilter.class);
    private String getCustomerLog(ServletRequest request)
    {
        //获取客户端请求头信息
        String header=((HttpServletRequest)request).getHeader("user-Agent");
        //创建处理客户端操作系统、浏览器及浏览器类型对象
        TerminateDevice terminate=new TerminateDevice(header);
        /*设置日志访问信息*/
        Log log=new Log();
        log.setBrowser(terminate.getBrowser());
        log.setBrowserInfo(terminate.getBrowserVersion());
        log.setOs(terminate.getOS());
        log.setIpAddress(request.getRemoteAddr());
        log.setProtocal(request.getProtocol());
        log.setPort(String.valueOf(request.getRemotePort()));
        log.setSessionId(((HttpServletRequest) request).getSession().getId());
        log.setUrl(((HttpServletRequest) request).getRequestURL().toString());
        if (((HttpServletRequest) request).getSession().getAttribute("user")!=null)
            log.setUser(((User)((HttpServletRequest) request).getSession().getAttribute("user")).getNumber());
```

```
            return log.toString();
    }
    @Override
    public void init(FilterConfig filterConfig) throws ServletException {

    }

    @Override
    public void doFilter(ServletRequest ServletRequest, ServletResponse ServletResponse, FilterChain filterChain) throws IOException, ServletException {
        //输出采集到的请求日志
        System.out.println(getCustomerLog(ServletRequest));
        //将日志信息写入日志文件
        logger.info(getCustomerLog(ServletRequest));
        //执行下一个过滤对象,如果是最后一个,则访问请求资源
        filterChain.doFilter(ServletRequest,ServletResponse);
    }
    @Override
    public void destroy() {
    }
}
```

其中,代码 logger.info(getCustomerLog(ServletRequest))实现了将采集到的日志信息写入日志文件中。

另外,代码 filterChain.doFilter(ServletRequest,ServletResponse)实现了客户端继续访问,如果后面还有其他的过滤器,则会转向下一个过滤器。

4. 向 Web 应用程序注册日志过滤器

在当前项目找到 web.xml 文件中的配置节点 Filter,具体配置如下:

```
<filter>
    <filter-name>logFilter</filter-name>
    <filter-class>chapter9.LogFilter</filter-class>
</filter>
```

5. 配置过滤器的过滤路径

向应用程序注册日志过滤器后,还是无法应用,必须在项目的 web.xml 中继续配置过滤器的映射路径,即哪些用户请求需要过滤。只要符合路径要求的都可以过滤,本案例中过滤了所有客户端请求,具体映射配置如下:

```
<filter-mapping>
    <filter-name>logFilter</filter-name>
    <url-pattern>/*</url-pattern>
</filter-mapping>
```

如果只是拦截部分,仅需要修改节点 url-pattern。

运行结果

案例启动后,控制台日志输出结果如图 9-5 所示。

当然,图 9-5 中结果每次都会不一样,它会根据访问路径、访问地址、访问时间及用户数据显示不同结果。

案例中的日志文件写入 d:\\jsplog\\info.log 中，如图 9-6 所示。

图 9-5　控制台日志

图 9-6　日志信息输出的文件

文件打开后，可以看到如图 9-7 所示的结果：

图 9-7　日志文件中的日志信息

9.3　单点登录及授权访问

单点登录在项目中的应用非常广泛，即同一个用户在同一时刻只能有一个用户账号在线，当一个在线用户登录成功后，在没有退出的前提下又用这个账号在另一处登录视为非法，必须先把之前登录的账号会话撤销，将最近登录设为有效登录。

案例描述

利用过滤器及监听器技术，实现单点登录。当前案例是一个综合性项目案例，功能较为复杂，具体功能如下：

- 实现登录用户的单点登录；

- 实现数据请求与服务器响应，编码格式统一为"utf-8"；
- 实现有效用户在线人数统计；
- 实现功能模块的权限访问。没有授权的无法访问，跳转登录页面。

案例分析

案例描述中的单点登录及授权访问，实际上是项目开发中的常规功能，这些功能在案例描述中分为单点登录、编码格式、在线人数统计及授权访问等 4 个，根据目前学习过的过滤器及监听器技术，单点登录及访问权限解决思路如图 9-8 所示。

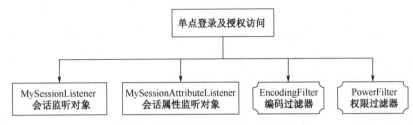

图 9-8　单点登录及访问权限解决思路

图 9-8 中，单点登录及授权访问分别有会话监听对象 MySessionListener、会话属性监听对象 MySessionAttributeListener、编码过滤器 EncodingFilter 和权限过滤器 PowerFilter 实现，它们的具体功能如下：

- MySessionListener 会话监听器对象实现在线人数统计功能；
- MySessionAttributeListener 会话属性监听对象实现单点登录功能；
- EncodingFilter 编码过滤器实现项目中的编码格式设置；
- PowerFilter 权限过滤器实现项目中的用户访问权限鉴别。

1. 会话监听对象 MySessionListener

会话监听对象 MySessionListener 负责实现在线用户人数统计功能，需要实现父类接口 HttpSessionListener，MySessionListener 类的设计见表 9-9。

表 9-9　MySessionListener 类的设计

类名称	MySessionListener	中文名称	会话监听对象	类型	实体类	父类	HttpSession-Listener
类描述	实现用户在线人数统计功能						
方法							
序号	方法名	参数		返回类型	功能说明		
1	sessionCreated	HttpSessionEvent event：会话事情		void	当一个新的会话对象存在时，在线人数自动增加 1		
2	sessionDestroyed	HttpSessionEvent event：会话事情		void	会话对象销毁，当执行该方法时意味着在线人数减少 1		

2. 会话属性监听对象 MySessionAttributeListener

会话属性监听对象 MySessionAttributeListener 实现单点用户登录功能，利用会话属性新增的特征，当一个用户登录成功后，要求将其登录信息写入 session 的"user"属性中，即 session.add（"user"，登录用户对象）。用户单点登录解决思路如图 9-9 所示。

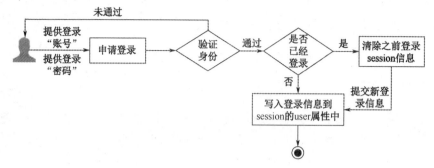

图 9-9　用户单点登录解决思路

图 9-9 中，用户登录时提供有效的账号及密码信息，通过系统身份验证，如验证通过，从全局变量"sessions"属性中查找已经登录的 session 对象，判断是否之前已经在其他地方登录过。如果登录过，则删除原来的登录信息，最后将最新的登录信息写入 session 的"user" 属性中，并将当前新 session 对象写入上下文全局对象 "sessions"属性中。

MySessionAttributeListener 类的设计见表 9-10。

表 9-10　MySessionAttributeListener 类的设计

类 名 称	MySessionAttributeListener	中文名称	会话属性监听对象	类　型	实体类	父类	HttpSessionAttributeListener
类 描 述	实现用户单点登录功能						
方　　法							
序　号	方 法 名	参　　数		返回类型		功能说明		
1	addSession	HttpSession sess：会话对象		void		新增会话属性，当新增"user"属性时，判断当前用户是否登录过，如果登录过，则注销之前登录过的 session 对象		
2	attributeAdded	HttpSessionBindingEvent event：会话绑定事件		void		新增会话属性		
3	attributeRemoved	HttpSessionBindingEvent event：会话绑定事件		void		删除会话属性		
4	attributeReplaced	HttpSessionBindingEvent event：会话绑定事件		void		会话属性值变更		

3. 编码过滤器 EncodingFilter

编码过滤器 EncodingFilter 实现用户请求数据格式及服务器端的数据响应格式。EncodingFilter 类的设计见表 9-11。

表 9-11　EncodingFilter 类的设计

类 名 称	EncodingFilter	中文名称	编码过滤器	类　型	实体类	父类	Filter
类 描 述	实现项目编码格式设置						
成员变量							
序　号	变量名		修 饰 词	类　　型		作　　用	
1	encoding		无	String		私有变量，存储编码格式	
2	config			FilterConfig		私有变量，过滤器参数变量	

续表

方法				
序号	方法名	参数	返回类型	功能说明
1	init	FilterConfig filterConfig：过滤配置参数对象	void	继承实现父接口中的方法，过滤器初始化时，此方法获取过滤器配置的编码格式参数值
2	doFilter	ServletRequest ServletRequest：页面请求对象；ServletResponse ServletResponse：页面响应对象；FilterChain filterChain：过滤链对象	void	继承实现父接口中的方法，核心内容，将获得的编码格式设置到客户端请求对象及服务器响应对象
3	destroy	无	void	继承实现父接口中的方法，此处的应用为空

4. 权限过滤器 PowerFilter

权限过滤器 PowerFilter 是根据用户请求访问的权限与已经存在的授权权限进行比对的，如果不存在，则取消访问请求，整个权限控制由权限接口、权限类及权限过滤器组成，权限访问控制设计思路如图 9-10 所示。

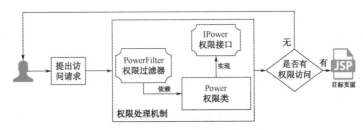

图 9-10　权限访问控制设计思路

图 9-10 中，用户提供访问服务器功能请求，请求转交给权限过滤器 PowerFilter，权限过滤器通过 Power 权限类查询并比对权限的有效性。如果验证通过，则允许用户访问目标页面，否则返回不允许访问。权限接口定义及权限类定义详见案例实现部分。PowerFilter 类的设计见表 9-12。

表 9-12　PowerFilter 类的设计

类名称	PowerFilter	中文名称	权限过滤器	类型	实体类	父类	Filter
类描述	实现用户访问权限过滤处理						
方法							
序号	方法名	参数		返回类型	功能说明		
1	init	FilterConfig filterConfig：过滤配置参数对象		void	继承实现父接口中的方法，过滤器初始化执行		
2	doFilter	ServletRequest ServletRequest：页面请求对象；ServletResponse ServletResponse：页面响应对象；FilterChain filterChain：过滤链对象		void	继承实现父接口中的方法，核心内容，实现对用户权限有效性验证		
3	destroy	无		void	继承实现父接口中的方法，此处的应用为空		

在权限过滤器中需要使用权限类 MyPower，其设计见表 9-13。

表 9-13 MyPower 类的设计

类名称	MyPower	中文名称		权限类	类型	实体类	父类	IPower，实现接口
类描述	实现对指定用户访问权限的设置							
成员变量								
序号	变量名	修饰词		类型		作用		
1	user	无		User		私有变量，权限用户		
方法								
序号	方法名	参数		返回类型		功能说明		
1	MyPower	User u:权限用户		MyPower		构造方法		
2	MyPower	无		MyPower		构造方法		
3	getUser	无		user		继承实现父接口中的方法，此处的应用为空		
4	setUser	User user：用户		void		设置授权用户		
5	getMyPower	无		List\<String\>		获取当前用户的系统权限		

案例实现

在案例分析中已将解决问题的思路进行了详细说明，功能实现需要按照以下几个步骤进行：

1. IPower 权限接口

IPower 权限接口中定义了获取用户权限集合的方法：

```java
package chapter9;
import business.entity.User;
import java.util.List;
/**
 * 用户权限接口
 * Created by wph-pc on 2018/11/8.
 */
public interface IPower {
    //获取当前用户的权限
    List<String> getMyPower();
}
```

2. MyPower 权限类定义

MyPower 权限类需要实现接口 IPower，案例中模拟的一些权限，实际应用中可以替换 getMyPower() 的方法体内容。例如可以从数据库中获取权限信息或从 NO-SQL 的 Redis 库中获取。案例中将一些静态配置的授权信息放置在资源属性文件"power.properties"中。属性文件"power.properties"中的属性值"anon"表示无须授权，因此，一些公共资源可以将其属性值设置为"anon"。"power.properties"案例中内容设置为：

```
chapter5/*=anon
chapter6/*=anon
chapter8/*=anon
css/*=anon
```

```
header/*=anon
images/*=anon
plugins/*=anon
script/*=anon
uploadfiles/*=anon
error.jsp=anon
index.jsp=anon
nopower.jsp=anon
logincheck=anon
user=anon
```

读者在配置上述权限属性值时，请根据项目实际情况进行设置。MyPower 类的定义如下：

```
package chapter9;

import business.controller.KesunSuperController;
import business.entity.User;

import java.io.IOException;
import java.util.ArrayList;
import java.util.Enumeration;
import java.util.List;
import java.util.Properties;

/**
 * 用户权限
 * Created by wph-pc on 2018/11/8.
 */
public class MyPower implements IPower {
    //权限用户
    private User user=null;
    public User getUser() {
        return user;
    }
    public MyPower(User u){
        this.user=u;
    }
    public MyPower(){}
    public void setUser(User user) {
        this.user = user;
    }

    @Override
    public List<String> getMyPower() {
        /*以下模拟数据，读者可以根据成员变量user从其他数据源中提取*/
        List<String> power=new ArrayList<>();
        power.add("nopower.jsp");
        power.add("codeimage.jsp");
        power.add("login.jsp");
        /*读取默认配置资源文件数据*/
        Properties props=new Properties();
        try {
            props.load(KesunSuperController.class.getClassLoader().
```

```
            getResourceAsStream
("power.properties"));
                } catch (IOException e) {
                    e.printStackTrace();
                }
                Enumeration properties=props.propertyNames();
                /*遍历权限属性集合，获取无须授权的配置信息*/
                while(properties.hasMoreElements()){
                    //获取当前正在遍历的属性名
                    String name=properties.nextElement().toString();
                    //获取当前正在遍历的属性值
                    String value=props.getProperty(name);
                    //如果属性值为"anon"表示无须授权，加入权限集合
                    if ("anon".equals(value)) power.add(name);
                }
                //如果用户是admin，则将power.jsp页面加入admin，此处为模拟数据
                if (getUser()!=null && getUser().getNumber().equals("admin"))
power.add("power.jsp");
                return power;
            }
        }
```

3. 设置权限过滤器 PowerFilter

权限过滤器 PowerFilter 实现权限过滤，对所有的访问判断其是否在指定用户权限集合中，如果不在，则无法访问，具体实现定义如下：

```
        package chapter9;
        import business.entity.User;
        import com.sun.deploy.net.HttpRequest;
        import javax.Servlet.*;
        import javax.Servlet.http.HttpServletRequest;
        import javax.Servlet.http.HttpServletResponse;
        import java.io.IOException;
        import java.util.List;
        /**
         * 权限清单过滤器
         * Created by wph-pc on 2018/11/8.
         */
        public class PowerFilter implements Filter {
            @Override
            public void init(FilterConfig filterConfig) throws ServletException {
            }
            @Override
            public void doFilter(ServletRequest ServletRequest, ServletResponse
ServletResponse, FilterChain filterChain) throws IOException, ServletException {
                HttpServletRequest request=(HttpServletRequest)ServletRequest;
                MyPower power=new MyPower();
                if (request.getSession().getAttribute("user")!=null)
                    power.setUser((User)request.getSession().getAttribute
("user"));
                List<String> lPowers=power.getMyPower();
                for(int i=0;lPowers!=null && lPowers.size()>0 && i<lPowers.size();
```

```java
i++){
            String uri=request.getRequestURL().toString();
            String tempPower=lPowers.get(i);
            if (tempPower.lastIndexOf("/*")>=0)
                tempPower=tempPower.substring(0,tempPower.lastIndexOf("/*"));
            if (uri.indexOf(tempPower)>=0) {
                filterChain.doFilter(ServletRequest,ServletResponse);
                return;
            }
        }
        ((HttpServletResponse)ServletResponse).sendRedirect(request.getContextPath() + "/chapter9/nopower.jsp");
    }
    @Override
    public void destroy() { }
}
```

权限过滤器的核心功能在 doFilter 方法中，实现权限鉴别，如果不在指定权限范围的内容，将会跳转到"nopower.jsp"页面。

开发完权限过滤器后，还无法直接发挥作用，需要在项目中的 web.xml 文件中注册过滤器及配置地址映射。

- 权限过滤器 PowerFilter 在 web.xml 中注册：

```xml
<filter>
    <filter-name>powerFilter</filter-name>
    <filter-class>chapter9.PowerFilter</filter-class>
</filter>
```

- 权限过滤器 PowerFilter 在 web.xml 中地址映射：

```xml
<filter-mapping>
    <filter-name>powerFilter</filter-name>
    <url-pattern>/chapter9/*</url-pattern>
</filter-mapping>
```

4. 编码过滤器 EncodingFilter

编码过滤器 EncodingFilter 实现项目中字符编码格式控制，主要对客户端请求对象 request 及服务器端响应对象 response 设置字符编码格式。EncodingFilter 的定义如下：

```java
package chapter9;
import javax.Servlet.*;
import java.io.IOException;
/**
 * 编码过滤器
 * Created by wph-pc on 2018/11/6.
 */
public class EncodingFilter implements Filter {
    //定义编码格式变量
    private String encoding = null;
    //定义过滤配置参数变量
    private FilterConfig config;
    @Override
```

```java
        public void init(FilterConfig filterConfig) throws ServletException {
            //获取配置参数
            this.config = filterConfig;
            //从web.xml配置文件中读取编码配置参数
            this.encoding = filterConfig.getInitParameter("encoding");

            System.out.println("过滤器名称："+filterConfig.getFilterName());
            System.out.println("encoding:"+filterConfig.getInitParameter("encoding"));
        }

        @Override
        public void doFilter(ServletRequest ServletRequest, ServletResponse ServletResponse, FilterChain filterChain) throws IOException, ServletException {
            if (ServletRequest.getCharacterEncoding()!=null &&
                    !encoding.equals(ServletRequest.getCharacterEncoding()) ||
                    ServletRequest.getCharacterEncoding()==null) {
                ServletRequest.setCharacterEncoding(this.encoding);
                ServletResponse.setCharacterEncoding(this.encoding);
            }
            filterChain.doFilter(ServletRequest,ServletResponse);
        }

        @Override
        public void destroy() {

        }
    }
```

开发 EncodingFilter 类后，必须在 web.xml 文件中注册，并进行访问地址映射。

■ EncodingFilter 类在 web.xml 文件中注册：

```xml
<filter>
    <filter-name>encoding</filter-name>
    <filter-class>chapter9.EncodingFilter</filter-class>
    <init-param>
        <param-name>encoding</param-name>
        <param-value>utf-8</param-value>
    </init-param>
</filter>
```

上面配置中，将编码格式配置为"utf-8"。

■ EncodingFilter 类在 web.xml 文件中地址映射配置：

```xml
<filter-mapping>
    <filter-name>encoding</filter-name>
    <url-pattern>/*</url-pattern>
</filter-mapping>
```

映射配置中，所有的地址请求都进行了编码格式的过滤。

5. 会话监听对象 MySessionListener

会话监听对象 MySessionListener 实现在线人数统计，当新建一个会话时，在线人数自动增加 1；一个会话销毁时，会减少 1 个在线人数。该类的定义如下：

```java
package chapter9;
import javax.Servlet.http.HttpSessionEvent;
import javax.Servlet.http.HttpSessionListener;
/**
 * 会话对象监听器
 * Created by wph-pc on 2018/11/7.
 */
public class MySessionListener implements HttpSessionListener {
    @Override
    public void sessionCreated(HttpSessionEvent httpSessionEvent) {
        if (httpSessionEvent.getSession().isNew()){
            if (httpSessionEvent.getSession().getServletContext().getAttribute("online")==null){
                httpSessionEvent.getSession().getServletContext().setAttribute("online",1);
            }else{
                int count=(int)httpSessionEvent.getSession().getServletContext().getAttribute("online");httpSessionEvent.getSession().getServletContext().setAttribute("online",count+1);
            }
        }
    }
    @Override
    public void sessionDestroyed(HttpSessionEvent httpSessionEvent) {
        if (httpSessionEvent.getSession().getServletContext().getAttribute("online")!=null){
            int count=(int)httpSessionEvent.getSession().getServletContext().getAttribute("online");
            if (count>0)
                httpSessionEvent.getSession().getServletContext().setAttribute("online",count-1);
        }
    }
}
```

监听器需要注册到应用程序中,通过配置当前项目中的 web.xml 监听节点,配置信息如下:

```xml
<listener>
    <listener-class>chapter9.MySessionListener</listener-class>
</listener>
```

6. 会话属性监听对象 MySessionAttributeListener

会话属性监听对象 MySessionAttributeListener 负责单点登录功能的实现,当用户登录后,判断当前在线用户是否存在,如果存在,则删除原来的登录信息。然后将登录的信息写入 session 的 "user" 属性中。该类的具体功能定义如下:

```java
package chapter9;
import business.entity.User;
import javax.Servlet.ServletContext;
import javax.Servlet.ServletContextListener;
import javax.Servlet.ServletRequestListener;
```

```java
import javax.Servlet.http.HttpSession;
import javax.Servlet.http.HttpSessionAttributeListener;
import javax.Servlet.http.HttpSessionBindingEvent;
import java.util.ArrayList;
import java.util.Iterator;
import java.util.List;

/**
 * Created by wph-pc on 2018/11/7.
 */
public class MySessionAttributeListener implements HttpSessionAttributeListener {
    /*单点登录,如果当前用户已经登录,则清除之前的登录session*/
    private void addSession(HttpSession sess){
        if (sess!=null &&
                sess.getAttribute("user")!=null){
            if (sess.getServletContext().getAttribute("sessions")!=null){
                //获取当前所有登录的用户session
                List<HttpSession> sessions=(List<HttpSession>)sess.getServletContext().getAttribute("sessions");
                /*判断当前session用户是否已经登录,如果已经登录,则注销已经登录的用户session*/
                Iterator<HttpSession> source=sessions.iterator();
                while(source.hasNext()){
                    HttpSession temp=source.next();
                    if (((User)temp.getAttribute("user")).getNumber().equals(((User)sess.getAttribute("user")).getNumber()) &&
                            !temp.getId().equals(sess.getId())){
                        /*注销原来的session,即temp*/
                        temp.invalidate();
                        //移除
                        sessions.remove(temp);
                        //退出循环
                        break;
                    }
                }
                //加入新的session
                sessions.add(sess);
            }
            else{
                List<HttpSession> sessions=new ArrayList<>();
                sessions.add(sess);
                sess.getServletContext().setAttribute("sessions", sessions);
            }
        }
    }
    @Override
    public void attributeAdded(HttpSessionBindingEvent httpSessionBindingEvent) {
```

```
        if ("user".equals(httpSessionBindingEvent.getName())){
            addSession(httpSessionBindingEvent.getSession());
        }
    }
    @Override
    public void attributeRemoved(HttpSessionBindingEvent httpSessionBindingEvent) {
    }
    @Override
    public void attributeReplaced(HttpSessionBindingEvent httpSessionBindingEvent) {
    }
}
```

MySessionAttributeListener 开发完成后，需要在当前项目的 web.xml 文件中注册监听该类，注册配置信息如下：

```
<listener>
    <listener-class>chapter9.MySessionAttributeListener</listener-class>
</listener>
```

7. 建立测试页面

在当前路径下，新建一个无须授权访问的页面"nopower.jsp"，其中不需要具体的页面内容，页面代码如下：

```
<%--
  Created by IntelliJ IDEA.
  User: wph-pc
  Date: 2018/11/8
  Time: 10:04
  To change this template use File | Settings | File Templates.
--%>
<%@ page contentType="text/html;charset=UTF-8" language="java" %>
<%@ include file="../header/init_bootstrap.jsp"%>
<html>
<head>
    <title>未授权页面</title>
</head>
<body>
<div class="well">
    您的操作未授权，联系管理员授权后，<a href="../chapter6/comprehensivecase/login.jsp">重新登录</a>。
</div>
</body>
</html>
```

再新建一个"power.jsp"页面，该页面需要授权才能访问，页面中不必作具体的内容和程序设置；其 HTML 编码如下：

```
<%--
  Created by IntelliJ IDEA.
  User: wph-pc
  Date: 2018/11/8
  Time: 10:00
  To change this template use File | Settings | File Templates.
```

```
--%>
<%@ page contentType="text/html;charset=UTF-8" language="java" %>
<%@include file="../header/init_bootstrap.jsp"%>
<html>
<head>
    <title>授权测试页面</title>
</head>
<body>
    <div class="well">此页面需要授权才能访问，能看到此页说明您已经获取了当前页面操作权</div>
</body>
</html>
```

运行结果

运行用户登录"login.jsp"页面，填写用户登录信息，身份验证通过后，页面跳转到"index.jsp"页面，如图 9-11 所示。

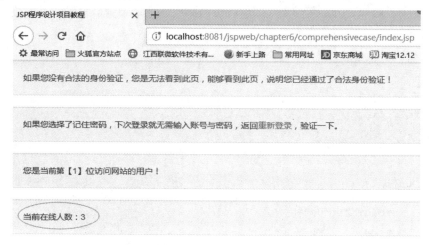

图 9-11 在线人数统计

为了验证单点登录，读者可以开启两个不同的浏览器代表不同的终端，启动登录页面，先登录成功第一个浏览器，不要关闭该浏览器；再用同一个用户信息在另外一个浏览器登录，登录成功后再去刷新第一个已经登录成功的页面，若发现该页面登录无效，则自动跳转到登录页面。

习题

在第 8 章中提到功能权限的基本信息维护，但是功能权限的应用需要将已经存在的权限授权给指定的系统角色或用户，当系统用户在使用系统中的功能时，系统需要判断当前用户是否拥有该项权限，如果没有，则跳转到指定未授权的页面，否则执行该项功能。本项目具体要求：

1．实现用户登录功能，除登录权限外，其他的权限使用都必须先登录；

2．如果没有用户、角色管理功能，请先实现用户和角色的新增、修改、删除及查询功能；

3．用户功能需要包括登录、退出、密码修改及角色分配；

4．继续使用第 8 章的系统权限管理功能，增加权限授权及权限使用；将已经存在的系统功能权限授权给指定的角色；

5．利用过滤器及监听器技术，实现权限使用的监督，未拥有指定权限则跳转到指定未授权页面；

6．开发项目中指定的功能操作页面。

第 10 章

JDBC 数据库开发

在信息系统开发中，数据库的地位非常重要，它起到数据存储及数据管理的作用。Java 语言不含数据库的功能，而 Java 开发的信息系统采集到的各种数据信息需要长久保存起来，Java 本身不具备这种功能，需要借助各种关系型数据库存储这些数据。目前的主流关系数据库包括：Oracle、DB、MySQL 和 SQL Server 等。Java 信息系统如何将产生的大量数据保存在这些数据库中及如何从这些数据库中提取 Java 信息系统所需要的数据，就是本章拟解决的核心问题，其中 JDBC 是关键技术。本章通过用户管理综合案例展示 JDBC 数据库开发技术的应用。

本章任务

（1）JDBC 开发配置；
（2）用户 CRUD 开发；
（3）用户登录与密码修改。

重点内容

（1）掌握 JDBC 开发配置，特别是连接缓冲池技术；
（2）掌握 JDBC 开发技术在项目中的应用。

难点内容

（1）JDBC 数据库缓冲池连接技术；
（2）JDBC 技术在项目中的应用。

10.1 JDBC 相关知识

JDBC（Java Database Connectivity），即 Java 数据库连接，用来建立 Java 应用程序与

关系数据库之间的连接，并执行 CRUD 相关操作。它是用于执行 SQL 语句的 Java API。CRUD 指的是 Create（新增）、Retrieve（查询）、Update（更新）和 Delete（删除）四种常规的数据库操作。Java 应用程序如何通过 JDBC 进行数据库 CRUD 操作？图 10-1 解析了数据库操作过程。

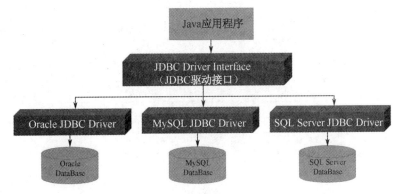

图 10-1 JDBC 数据库访问示意图

图 10-1 中，Java 应用程序通过 JDBC 驱动接口，访问不同类型的数据库，不同类型的数据库需要有专用的 JDBC Driver，例如：Oracle 数据库需要使用 Oracle JDBC Driver 驱动，MySQL 数据库需要 MySQL JDBC Driver 驱动，SQL Server 数据库需要 SQL Server JDBC Driver 驱动。

10.1.1 JDBC 核心类

JDBC 驱动接口提供了一系列有关数据库的操作类，这些类的作用与 Java、数据库之间的关系如图 10-2 所示。

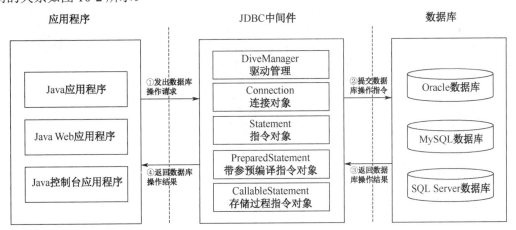

图 10-2 JDBC 类与 Java、数据库中间关系

图 10-2 中，Java 应用程序可以是桌面应用程序、Java Web 应用程序或 Java 控制台应用程序，应用程序通过向 JDBC 发出数据库指令请求，JDBC 中间件接收到操作指令后，进行数据库连接。如果连接不上，终止操作，并出现异常信息；如果连接正常，则通过相关指令对象，提交给指定的数据库。数据库执行结束后，再将执行结果返回给 JDBC 中间件，

最后返回给 Java 应用程序。

1. DriverManager

DriverManager 是管理 JDBC 驱动的服务类，主要通过它获取 Connection 数据库连接对象。该类的定义见表 10-1。

表 10-1 DriverManager 类的定义

类 名 称	DriverManager	中文名称	驱动管理	类 型	实体类
类 描 述	提供驱动管理相关功能				
序 号	方法名	参 数	返回类型	功能说明	
1	DriverManager	无	DriverManager	私有的构造方法	
2	getConnection	String url：数据库连接地址，包括端口号和数据库名称；String user：数据库访问用户；String password：数据库访问密码	Connection	获取数据库连接对象，静态的，并且是同步操作	

下面是几种常用数据库连接驱动配置：

- Oracle 数据库连接驱动配置

```
驱动程序包名：ojdbc14.jar
驱动程序类名：Oracle.jdbc.driver.OracleDriver
JDBC URL：
jdbc:oracle:thin:@//<主机名或IP地址>:<端口号>/服务名称
或
jdbc:oracle:thin:@<主机名或IP地址>:<端口号>:<服务SID>
```

- MySQL 数据库连接驱动配置

```
驱动程序包名：MySQL-connector-Java-版本号-bin.jar
驱动程序类名:com.mysql.jdbc.Driver
JDBC URL:jdbc:mysql://<主机名或IP地址>:<端口号>/<数据库名称>
默认端口3306,如果服务器使用默认端口则port可以省略
说明：MySQL Connector/J Driver允许在URL中添加额外的连接属性jdbc:mysql:// <主机名或IP地址>:<端口号>/<数据库名称>?属性名1=属性值1&属性名2=属性值2&...
```

- SQL Server 2005 版本驱动连接配置

```
驱动程序包名：sqljdbc.jar
驱动程序类名:com.microsoft.sqlserver.jdbc.SQLServerDriver
JDBC URL:jdbc:sqlserver:// <主机名或IP地址>:<端口号>
默认端口1433,如果服务器使用默认端口则port可以省略
```

2. Connection

Connection 用于实现对指定的数据库进行连接，连接期间可以执行数据库操作指令及获取数据库执行结果。Connection 接口描述见表 10-2。

表 10-2　Connection 接口描述

类 名 称	Connection	中文名称	连接对象	类　型	接口
类 描 述	提供数据库连接对象并进行数据库相关功能的定义				
序　号	方法名	参　数	返回类型	功能说明	
1	createStatement	无	Statement	获取一个数据库操作指令对象 Statement	
2	prepareStatement	Sring sql：数据库操作指令字符串	PreparedStatement	获取预编译的 Statement 数据库指定对象	
3	prepareCall	Sring sql：指令字符串，其中可以含有一个或多个 "？"	CallableStatement	返回一个 CallableStatement 对象，用于存储过程的调用	
4	setSavepoint	String name：设置事务保存点	Savepoint	设置事务保存点	
5	setTransactionIsolation	int level：级别	void	设置事务隔离级别	
6	rollback	无	void	回滚	
7	rollback	Savepoint savepoint：事务回滚点	void	回滚到指定事务点 savepoint	
8	commit	无	void	事务提交	
9	close	无	void	关闭连接	

3．Statement

Statement 是用于执行 SQL 语句的 API 接口，该对象可以执行数据定义（DDL）语句、数据控制（DCL）语句，也可以执行数据操作（DML）等，还可以执行 SQL 查询语句，以结果集的形式返回。Statement 接口描述见表 10-3。

表 10-3　Statement 接口描述

类 名 称	Statement	中文名称	指令对象	类　型	接口
类 描 述	提供对数据库 CRUD 的功能定义				
序　号	方法名	参　数	返回类型	功能说明	
1	executeUpdate	String sql：数据库字符串指令	int	用于执行 DML 语句，并返回受影响的行数；该方法也可以执行 DDL，执行 DDL 返回 0。	
2	executeQuery	String sql：数据库查询字符串指令	ResultSet	用于执行查询语句，并返回查询结果对应的 ResultSet 对象	
3	execute	String sql：数据库字符串指令	boolean	方法可以执行任何 SQL 语句，如果执行后第一个结果是 ResultSet 对象，则返回 true；如果执行后第一个结果为受影响的行数或没有任何结果，则返回 false	

4．PreparedStatement 指令接口

PreparedStatement 接口是预编译的 Statement 对象，PreparedStatement 是 Statement 的子接口，它允许数据库预编译 SQL（通常指带参数 SQL）语句，之后的使用仅需改变 SQL 命令参数，避免数据库每次都编译 SQL 语句，有效提升了数据库的操作效率。而相对于 Statement 而言，使用 PreparedStatement 执行 SQL 语句时，由于之前已经预编译了 SQL 语

句，因此无须重新传入 SQL 语句，但是 PreparedStatement 需要为编译的 SQL 语句传入参数值。

PreparedStatement 对象用于发送带有一个或多个输入参数（IN 参数）的 SQL 语句。PreparedStatement 拥有一组方法，用于设置 IN 参数的值。执行语句时，这些 IN 参数将被发送到数据库中；传值的方法类似：

```
void setInt(int index, int value)
```

上面方法表示将整型值 value 传给 PreparedStatement 的第 index 个参数。根据该方法传入的参数值的类型不同，需要使用不同的方法。传入的值类型根据传入的 SQL 语句参数而定，例如：setFloat、setDouble、setString、setLong、setObject 等。

5. CallableStatement

CallableStatement 是 PreparedSatement 的子接口，用于执行 SQL 储存过程 CallableStatement 对象提供了可以处理 IN 参数的方法，还增加了用于处理 OUT 参数和 INPUT 参数的方法。本章节没有涉及此案例，故不做详细阐述，如需详细资料，读者可以自行查阅。

6. ResultSet

ResultSet 用于存放从数据库中查询的数据，并以二维表格形式返回。它是一个接口，提供在线阅读查询的数据，一旦数据库连接断开，就无法继续读取数据。ResultSet 接口描述见表 10-4。

表 10-4 ResultSet 接口描述

类名称	ResultSet	中文名称	结果集	类型	接口
类描述	提供读取结果集中的相应功能定义				
序号	方法名	参数	返回类型	功能说明	
1	close	无	void	释放、关闭 ResultSet 对象	
2	absolute	int row：行数	boolean	将结果集移动到第 row 行。如果移动到的记录指针指向一条有效记录，则该方法返回 true	
3	beforeFisrt	无	void	将 ResultSet 的记录指针定位到首行之前，这是 ResultSet 结果集记录指针的初始状态，记录指针的起始位置位于第一行之前	
4	first	无	boolean	将 ResultSet 的记录指针定位到首行。如果移动后的记录指针指向一条有效记录，则该方法返回 true	
5	previous	无	boolean	将 ResultSet 的记录指针定位到上一行，如果移动后的记录指针指向一条有效记录，则该方法返回 true	
6	next	无	boolean	将 ResultSet 的记录指针定位到下一行。如果移动后的记录指针指向一条有效记录，则返回 true	
7	last	无	boolean	将 ResultSet 的记录指针定位到最后一行。如果移动后的记录指针指向一条有效记录，则返回 true	
8	afterLast	无	boolean	将 ResultSet 的记录指针定位到最后一行之后	

10.1.2 JDBC 连接池配置

Java 应用程序与数据库之间的访问，需要利用 JDBC 实现对目标数据库的连接，而连接数据库在整个数据库操作中占用了大量的资源，非常耗时。如果多个终端对数据库同时进行连接，数据库的操作性能会大大下降，严重时会导致数据库连接阻塞从而无法访问。为了较好地管理好数据库连接，一般会采用连接池的方式进行管理，本节介绍的连接池是由 Apache 组织开发的，其代码开源。该连接池负责管理数据库的连接，当用户不需要连接时会自动收回。

1. DBCP 相关知识

DBCP（DataBase Connection Pool）数据库连接池，是 Java 数据库连接池的一种，由 Apache 开发，通过数据库连接池，可以让程序自动管理数据库连接的释放和断开。

DBCP 是 Apache 上的一个 Java 连接池项目，也是 Tomcat 使用的连接池组件。单独使用 DBCP 需要 2 个包：commons-dbcp.jar 和 commons-pool.jar。通过 DBCP 解决因建立数据库连接而产生耗时、耗资源的问题，连接池预先同数据库建立一些连接，放在内存中，应用程序需要建立数据库连接时，直接到连接池中申请一个即可，使用结束后再归还到连接池中。

2. DBCP 连接池配置步骤

DBCP 将数据库连接驱动进行了封装，数据库的连接对象不再直接通过 DriverManager 对象创建，而是通过 DBCP 获取连接对象 Connection。JDBC 连接池配置步骤如图 10-3 所示。

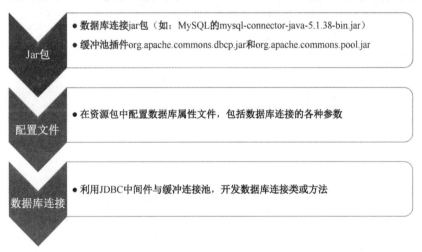

图 10-3　JDBC 连接池配置步骤

首先，下载必需的 jar 包，包括用于连接数据库的 JDBC Driver 驱动 jar 包；连接池包 org.apache.commons.dbcp.jar 和 org.apache.commons.pool.jar，其中 dbcp.jar 包需要依赖 pool.jar 包。

其次，配置连接池所需要的各种配置资源，一般，配置连接池所需的资源参数放置在当前项目资源包中的属性文件中，方便连接池读取。

最后，开发带连接池功能的连接方法，该功能本书放在 DBHelper 类中实现，后续有详细的讲解。

连接池所需的各种数据库连接参数，本章案例写在资源文件 db.properties 中，配置内容如下：

```
jdbc.driver=com.mysql.jdbc.Driver
jdbc.url=jdbc\:mysql\://服务器IP地址或名称\:3306/数据库名称
jdbc.user=root
jdbc.password=root
dataSource.initialSize=10
dataSource.maxIdle=20
dataSource.minIdle=5
dataSource.maxActive=5000
dataSource.maxWait=1000
```

在上面的资源配置中，连接的是 MySQL 数据库，配置了 JDBC 数据库连接的必要参数，如：jdbc.driver 连接驱动名称，jdbc.url 数据库连接路径，jdbc.user 数据库访问用户，jdbc.password 数据库访问用户密码。同时，还配置了 DBCP 连接池所需的其他相关参数，例如：dataSource.initialSize 初始化连接数，dataSource.maxIdle 最大闲置数，dataSource.minIdel 最小闲置数，dataSource.maxActive 最大连接数，dataSource.maxWait 最大等待数。

10.1.3 单例模式 DBHelper 类

Java 应用程序对数据库进行指令操作，需要借助 JDBC 中的各种数据库操作对象完成，为了更好地将 DBCP 与 JDBC 有效结合，更加方便实现对数据库进行指令操作，将对数据库的操作 CRUD 集成在一块，DBHelper 类就承担了这个角色。

DBHelper 类实现了数据库访问单例模式，单例模式保证了在应用程序运行期间，DBHelper 的对象只有一个。DBHelper 类集成了对数据库的连接、新增、修改、删除及查询等操作功能，它在三层架构设计中的数据访问层作用如图 10-4 所示。

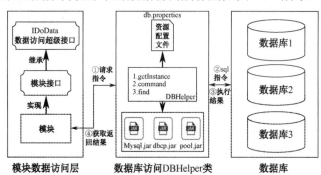

图 10-4 DBHelper 类在数据访问层中作用

图 10-4 中，模块数据访问层中"模块接口"需要继承"IDoData"接口，"模块"再实现"模块接口"，"模块"中的任何有关数据库操作指令都将发送到"数据库访问 DBHelper 类"中。DBHelper 类对外提供了获取单例对象静态方法 getInstance，执行数据库的 CUD 操作的 command 方法，执行数据查找的方法 find。DBHelper 功能的实现需要依赖 DriverManager 驱动包，连接池 dbcp.jar 和 pool.jar 包，以及数据库参数配置资源文件 db.properties。

1. DBHelper

DBHelper 类描述见表 10-5。

表 10-5 DBHelper 类描述

类 名 称	DBHelper	中文名称	数据库访问类	类 型	实体类
类 描 述	提供对数据库的 CRUD 基本操作定义				
变量定义					
序 号	变量名	修 饰 词	类 型	说 明	
1	sql	private static	DBHelper	DBHelper 单例对象	
2	dataSource	private	BasicDataSource	连接池数据源	
方法定义					
序 号	方法名	参 数	返回类型	功能说明	
1	getInstance	无	DBHelper	获取 DBHelper 单例对象,类方法	
2	init	无	void	私有方法,实现对 dataSource 数据源对象进行初始化	
3	command	String sql:增、删、改数据库指令字符串	int	实例方法,根据参数 sql 实现对数据库进行操作,通常执行带有影响行结果的操作,返回结果为影响行的行数	
4	convertList	ResultSet rs:数据结果集	List<Map<String, Object>>	私有方法,将参数 rs 转换成 List<Map<String,Object>>结构,提供 ResultSet 结果集离线阅读	
5	find	String sql:查询指令	List<Map<String, Object>>	实例方法,提供了根据 sql 查询指令进行数据库查询操作,并返回 List 结果集	

2. DBHelper 类的详细代码实现

```
package chapter10;
import org.apache.commons.dbcp.BasicDataSource;
import java.sql.*;
import java.util.*;
/**
 * 数据库访问类,通过连接池技术实现对数据库连接的自动管理
 * 实现了数据库的CRUD操作
 * Created by wph-pc on 2018/11/28.
 */
public class DBHelper {
    //静态DBHelper对象变量
    private static DBHelper sql=null;
    //定义缓冲池数据源
    private BasicDataSource dataSource=null;
    /*
     * 构造方法私有化,DBHelper单例模式的前提条件
     * */
    private DBHelper(){}
    /*
```

```java
 * 类方法,实现DBHelper对象单例实例化
 * @return 返回DBHelper对象
 * */
public static DBHelper getInstance(){
    if (sql==null)
        sql=new DBHelper();
    return sql;
}
/*连接数据库数据源初始化,需要common-pool和dbcp中间件支持*/
private void init(){
    //创建属性读取对象
    Properties dbprops=new Properties();
    //配置文件可以根据实际修改
    try {
        dbprops.load(DBHelper.class.getClassLoader().getResourceAsStream("db.properties"));
    } catch (Exception e) {
        e.printStackTrace();
    }
    try {
        /*读取属性文件中JDBC数据库连接参数的配置信息*/
        String driverClassName=dbprops.getProperty("jdbc.driver");
        String url=dbprops.getProperty("jdbc.url");
        String username=dbprops.getProperty("jdbc.user");
        String password=dbprops.getProperty("jdbc.password");
        /*读取属性文件中连接池配置信息*/
        String initialSize=dbprops.getProperty("dataSource.initialSize");
        String minIdle=dbprops.getProperty("dataSource.minIdle");
        String maxIdle=dbprops.getProperty("dataSource.maxIdle");
        String maxWait=dbprops.getProperty("dataSource.maxWait");
        String maxActive=dbprops.getProperty("dataSource.maxActive");

        //数据源对象实例化
        dataSource =new BasicDataSource();
        /*配置数据源相关参数信息*/
        dataSource.setDriverClassName(driverClassName);
        dataSource.setUrl(url);
        dataSource.setUsername(username);
        dataSource.setPassword(password);
        //配置初始化连接数
        if(initialSize!=null){
            dataSource.setInitialSize(Integer.parseInt (initialSize));
        }
        //配置最小空闲连接
        if(minIdle!=null){
            dataSource.setMinIdle(Integer.parseInt(minIdle));
        }
        //配置最大空闲连接
        if(maxIdle!=null){
```

```java
        dataSource.setMaxIdle(Integer.parseInt(maxIdle));
    }
    //配置超时回收时间（以毫秒为单位）
    if(maxWait!=null){
        dataSource.setMaxWait(Long.parseLong(maxWait));
    }
    //配置最大连接数
    if(maxActive!=null){
        if(!maxActive.trim().equals("0")){
            dataSource.setMaxActive(Integer.parseInt(maxActive));
        }
    }

    } catch (Exception e) {
        e.printStackTrace();
    }
}
/*
 * 数据库的增、删、改操作
 * @param sql:增、删、改数据库字符串指令
 * @return 返回操作影响行的行数
 * */
public int command(String sql){
    //判断数据源是否为空
    if (dataSource==null)
        init();
    if (dataSource==null) return -1;
    //定义连接对象变量
    Connection conn=null;
    //获取数据源中的连接对象
    try {
        //获取连接对象
        conn=dataSource.getConnection();
        //获取指令对象
        Statement state=conn.createStatement();
        //执行SQL指令
        return state.executeUpdate(sql);
    } catch (SQLException e) {
        e.printStackTrace();
        return -1;
    }finally {
        if (conn!=null)
            try {
                conn.close();
            } catch (SQLException e) {
                e.printStackTrace();
            }
    }
}
/*
 * 实现将结果集rs转换成List结果
```

```java
     * @param rs 结果集
     * @return 如果转换成功,返回List结果,否则返回null
     */
    private List<Map<String, Object>> convertList(ResultSet rs) throws Exception {
        //判断参数rs有效性
        if(rs == null) {
            return null;
        }
        //获取rs参数数据元结构
        ResultSetMetaData rsMetaData = rs.getMetaData();
        //获取数据源rs中的列数
        int columnCount = rsMetaData.getColumnCount();
        //创建List对象
        List<Map<String, Object>> dataList = new ArrayList<Map<String, Object>>();
        /*遍历rs结果集,并将遍历到的结果转换成List*/
        while (rs.next()) {
            //定义存放当前行中所有列的数据集合
            Map<String, Object> dataMap = new HashMap<String, Object>();
            //处理当前遍历行所在的所有列数据,列索引从1开始
            for (int i = 0; i < columnCount; i++) {
                dataMap.put(rsMetaData.getColumnName(i+1), rs.getObject(i+1));
            }
            //将遍历到当前行的所有数据放入dataList中
            dataList.add(dataMap);
        }
        return dataList;
    }
    /*
     * 根据参数sql从数据库中查询,并以List结果返回
     * @param sql:数据库查询指令
     * @return 返回查询结果,如果存在,
     *         以List返回,否则返回null
     */
    public List<Map<String,Object>> find(String sql){
        //判断数据源是否为空
        if (dataSource==null)
            init();
        if (dataSource==null) return null;
        //定义连接对象变量
        Connection conn=null;
        //获取数据源中的连接对象
        try {
            //获取连接对象
            conn=dataSource.getConnection();
            //获取指令对象
            Statement state=conn.createStatement();
            //执行SQL指令
```

```
            ResultSet rs=state.executeQuery(sql);
            if (rs!=null)
            try {
                List<Map<String,Object>> result=convertList(rs);
                rs.close();
                return result;
            } catch (Exception e) {
                e.printStackTrace();
                return null;
            }
            return null;

        } catch (SQLException e) {
            e.printStackTrace();
            return null;
        }finally {
            if (conn!=null)
            try {
                conn.close();
            } catch (SQLException e) {
                e.printStackTrace();
            }
        }
    }
}
```

10.2 用户 CRUD 开发

用户 CRUD 开发是指实现用户信息新增、查找、更新及删除操作。用户基本信息在任何信息系统中都存在，用户信息新增可以通过后台统一添加，也可以通过用户自己注册添加。用户信息修改和删除主要是对已经存在的用户信息进行修改或删除。用户查询则是对已经存在的用户信息进行查找操作。

案例描述

利用 JSP 技术、JDBC 技术、DBCP 连接池技术及 Ajax 技术实现对用户信息 CRUD 操作，具体要求如下：

- 用户（User）的基本信息包括账号（id）、昵称(name)、密码（password）、创建时间（createDate）和状态（status）等；
- 用户的账号信息不能重复；
- 数据库技术采用 JDBC，且需要使用连接池进行数据库的连接管理；
- 数据库采用 MySQL，表名称为 Users；
- 用户 CRUD 操作采用三层架构技术，并且在三层架构中表现层采用 MVC 机制；
- 页面数据交互技术采用 Ajax 技术；

- 数据查找支持按照账号且模糊查询。

案例分析

根据案例描述，实现对用户基本信息进行增、删、改、查的相关操作，不仅是基本实现，而且还需要采用三层架构及 MVC 模式；另外，要求采用 JDBC 技术及连接池技术解决数据库的连接及相关指令操作，DBHelper 类解决了数据库操作中 CRUD 问题。用户 CRUD 解决方案如图 10-5 所示。

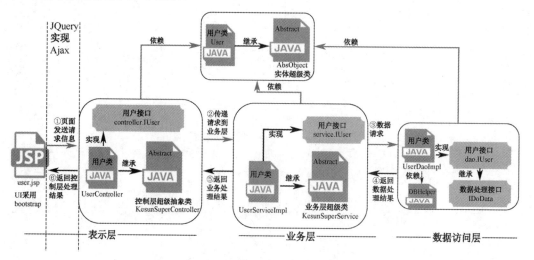

图 10-5　用户 CRUD 解决方案

图 10-5 中，对三层架构进行了详细说明，并提供了更为详细的设计：

- 数据访问层用户类 UserDaoImpl 需要依赖 DBHelper 类，通过 DBHelper 类的协助，实现用户数据的 CRUD 操作；
- 为了提升用户 UI 体验，页面 UI 实现采用 BootStrap 技术，如果读者对该技术不熟悉，可使用基本的 HTML 技术解决；
- 针对页面与 Servlet 控制层的数据交互，为了避免频繁的跳转，采用了 JQuery 的 Ajax 异步技术，这也是项目中常用的技术。

按照图 10-5 的设计思路，需要开发完成三层架构中的表示层、业务层及数据访问层中的相关模块。

页面与控制层进行数据交互时采用了 JQuery 的 Ajax 数据交互技术，数据标准采用 JSON 格式，在进行用户信息增、删、改操作时，需要将客户端请求的数据在控制层（Servlet 类中）转换成用户类 User，页面数据转成目标 JavaBean 对象如图 10-6 所示。

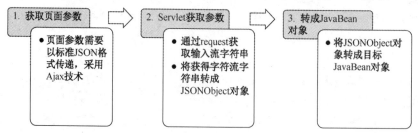

图 10-6　页面数据转成目标 JavaBean 对象

图 10-6 中，JSP 页面的数据采用 JSON 格式传递到控制层 Servlet 对象，Servlet 对象通过 request 客户端请求对象获取输入流对象，再将其转换成字符串，然后通过 fastjson 中间件，转换成 JSONObject 对象，这时就可以将 JSONObject 对象转成目标 JavaBean 对象或 Map 结构的数据。关于如何将页面端参数传递到控制层，并转成具体的 JavaBean 对象，本章专门设计了一个工具类 JSONTool 实现，其提供了相应的转换机制，JSONTool 类的设计见表 10-6。

表 10-6 JSONTool 类的设计

类 名 称	JSONTool	中文名称	JSON 工具类	类 型	实体类	父类	无
类 描 述	实现对 JSON 格式的处理，需要借助 fastjson 中间件							
方 法							
序 号	方 法 名	参 数		返回类型			功能说明	
1	GetPostData	InputStream in：输入流对象； int size：字符大小； String charset：字符格式名称		String			类方法，将输入流对象 in 中的字符大小及格式转成字符串	
2	GetRequestJSON	HttpServletRequest request： 请求对象		JSONObject			类方法，将 request 对象请求参数转换成 JSONObject 对象	
3	GetDate	String value： 字符串日期值参数		Date			类方法，将字符串 value 日期值转换成目标对象 Date	
4	JSONObjectToJavaBean	JSONObject source： json 数据源； Class<T> javaBean：目标对象		T			类方法，将参数 json 转换成 T 类型 JavaBean 对象	

JSONTool 类具体实现代码定义：

```
package util;

import com.alibaba.fastjson.JSONArray;
import com.alibaba.fastjson.JSONException;
import com.alibaba.fastjson.JSONObject;

import javax.Servlet.http.HttpServletRequest;
import javax.Servlet.http.HttpServletResponse;
import java.io.IOException;
import java.io.InputStream;
import java.lang.reflect.InvocationTargetException;
import java.lang.reflect.Method;
import java.lang.reflect.ParameterizedType;
import java.text.DateFormat;
import java.text.ParseException;
import java.text.SimpleDateFormat;
import java.util.Date;
import java.util.List;

/**
 * JSON数据格式处理工具
 * Created by wph-pc on 2018/12/6.
```

```java
        */
     public class JSONTool {
         /*
         *获取输入流in对象的字符串值
         * @param in:客户端输入流对象
         * @param size:输入流大小
         * @param charset:字符集名称
         * @return 返回字符流字符串内容
         **/
         public static String GetPostData(InputStream in, int size, String charset) {
             if (in != null && size > 0) {
                 byte[] buf = new byte[size];
                 try {
                     in.read(buf);
                     if (charset == null || charset.length() == 0)
                         return new String(buf,"utf-8");
                     else {
                         return new String(buf, charset);
                     }
                 } catch (IOException e) {
                     e.printStackTrace();
                 }
             }
             return null;
         }
         /*
         * 获取客户端请求参数,客户端的数据必须为JSON对象,返回JSON对象
         * @param request:客户端请求对象
         * @return 返回客户端请求JSON格式数据
         * */
         public static JSONObject GetRequestJSON(HttpServletRequest request)
         {
             try
             {
                 //获取客户端request输入流字符串
                 String json= GetPostData(request.getInputStream(), request.getContentLength(), "utf-8");
                 //将输入流字符串转换成JSON对象
                 return JSONObject.parseObject(json);
             }catch(IOException e)
             {
                 return null;
             }

         }
         /*
         * 将日期字符串value值转换成日期对象
         * @param value:日期字符串
         * @return :返回日期Date对象,只含有年、月、日
         * */
```

```java
public static Date GetDate(String value)
{
    //定义日期格式对象
    DateFormat df=new SimpleDateFormat("yyyy-MM-dd");
    try {
        //返回转换后的日期对象
        return df.parse(value);
    } catch (ParseException e) {
        // TODO Auto-generated catch block
        e.printStackTrace();
        return new Date();
    }
}
/*
 *将JSON数据转换成JavaBean对象
 * @param source:JSON数据源
 * @param javaBean:转换的目标对象
 * @return 返回T类型JavaBean对象
 * */
public static <T> T JSONObjectToJavaBean(JSONObject source,Class<T> javaBean)
{
    if (source==null) return null;
    Method[] beanMethods = javaBean.getMethods();//获取指定对象所有的JavaBean方法
    T tempBean = null;
    try {
        tempBean = javaBean.newInstance();
    } catch (Exception e){
        e.printStackTrace();
        return null;
    }
    for (Method method : beanMethods)
    {
        String field = method.getName();
        if (field.indexOf("set")<0) continue;

        String oldField=field.substring(3);
        field = field.substring(3);

        field = field.substring(0,1).toLowerCase() + field.substring(1);
        if (source.containsKey(field)==false)
            continue;
        else {
            try {
                if (source.get(field) instanceof JSONObject)
                {
                    method.invoke(tempBean, new Object[] {JSONObjectToJavaBean(source.getJSONObject(field), method.getParameterTypes()[0])});
```

```java
                    } else if (source.get(field) instanceof JSONArray)
                    {
                        List<Object> lTemp=new java.util.ArrayList<Object>();
                        JSONArray jArray=source.getJSONArray(field);
                        for(int i=0;i<jArray.size();i++)
                        {
                            ParameterizedType pt=null;
                            try {
                                pt = (ParameterizedType)tempBean.getClass().getMethod("get"+oldField).getGenericReturnType();
                            } catch (SecurityException e) {
                                // TODO Auto-generated catch block
                                e.printStackTrace();
                            } catch (NoSuchMethodException e) {
                                // TODO Auto-generated catch block
                                e.printStackTrace();
                            }
                            lTemp.add(JSONObjectToJavaBean(jArray.getJSONObject(i),(Class)pt.getActualTypeArguments()[0]));
                        }
                        method.invoke(tempBean, new Object[] {lTemp});//
                    }
                    else {
                        if (source.get(field)==null) continue;
                        if (method.getGenericParameterTypes()[0].toString().equals("class java.util.Date"))//处理日期格式
                        {
                            method.invoke(tempBean, new Object[] {GetDate(source.get(field).toString())});
                        }else if (method.getGenericParameterTypes()[0].toString().equals("int"))
                        {
                            method.invoke(tempBean, new Object[] {Integer.parseInt(source.get(field).toString())});
                        } else if (method.getGenericParameterTypes()[0].toString().equals("float"))
                        {
                            method.invoke(tempBean, new Object[] {Float.parseFloat(source.get(field).toString())});
                        }
                        else
                            method.invoke(tempBean, new Object[] {source.get(field)});
                    }
                } catch (IllegalArgumentException e) {
                    // TODO Auto-generated catch block
                    System.out.println("异常属性【"+field+"】"+method.getGenericParameterTypes()[0].toString());
                    e.printStackTrace();
                } catch (IllegalAccessException e) {
```

```
                    // TODO Auto-generated catch block
                    e.printStackTrace();
                } catch (InvocationTargetException e) {
                    // TODO Auto-generated catch block
                    e.printStackTrace();
                } catch (JSONException e) {
                    System.out.println("异常JSON属性【"+field+"】");
                    // TODO Auto-generated catch block
                    e.printStackTrace();
                }
            }
        }
        return tempBean;
    }
}
```

最后在 MySQL 中新建数据库，并在数据库中新建存放用户数据的表格 users，users 用户表定义格式见表 10-7。

表 10-7 users 用户表定义格式

序 号	列 名 称	类 型	长 度	是否主键	是否允许空	默 认 值	说 明
1	id	字符型	20	是	否	无	账号
2	name	字符型	20	否	否	无	昵称
3	pwd	字符型	40	否	否	无	密码
4	createDate	日期型		否	是	系统时间	创建日期
5	status	字符型	10	否	是	无	状态

案例实现

1. 数据访问层

数据访问层由用户类 UserDaoImpl、用户接口 dao.IUser、数据处理接口 IDoData 及 DBHelper 类组成。

（1）UserDaoImpl 类

UserDaoImpl 类实现了 dao.IUser 接口及用户密码修改及登录功能，具体实现如下：

```java
package business.dao.impl;

import business.dao.IUser;
import business.entity.User;
import business.entity.AbsObject;

import java.sql.DriverManager;
import java.util.*;

/**
 * 用户操作数据访问层模拟类
 * Created by wph-pc on 2018/10/31.
 */
public class UserDaoImpl implements IUser {
```

```java
/*将参数obj转换成User对象,如果不符合条件,返回null,否则返回User*/
private User getUser(AbsObject obj){
    //判断obj是否是用户类型
    if (obj==null || obj instanceof User==false) return null;
    //将obj转换成目标类型
    User u=(User)obj;
    return u;
}
@Override
public int add(AbsObject obj) {
    //将obj转换成目标类型
    User u=getUser(obj);
    //判断u是否为null
    if(u==null) return 0;

    u.setCreateDate(new Date());
    //将u对象转换成SQL字符串新增指令
    String sql="insert into users "+
        "(id,name,pwd,createDate,status) values "+
        "('"+u.getId()+"','"+u.getName()+"','"+
        util.MyMD5.GetMyUpperMD5(u.getPassword(),u.getId())+
        "',now(),'"+u.getStatus()+"')";
    //创建数据库访问对象
    chapter10.DBHelper command=chapter10.DBHelper.getInstance();
    //执行用户新增操作
    return command.command(sql);
}

@Override
public int edit(AbsObject obj) {
    //将obj转换成目标类型
    User u=getUser(obj);
    //判断u是否为null
    if(u==null) return 0;
    //将u对象转换成用户更新的SQL指令
    String cmd="update users set name='"+u.getName()+
        "',createDate=now(),status='"+u.getStatus()+
        "' where id='"+u.getId()+"'";
    //实例化数据访问对象DBHelper
    chapter10.DBHelper sql= chapter10.DBHelper.getInstance();
    //执行用户修改操作
    return sql.command(cmd);
}

@Override
public int del(AbsObject obj) {
    //将obj转换成目标类型
    User u=getUser(obj);
    //判断u是否为null
    if(u==null) return 0;
    //将对象u转换成SQL删除指令
```

```java
            String cmd="delete from users where id='"+u.getId()+"'";
            //创建数据库DBHelper访问对象
            chapter10.DBHelper sql=chapter10.DBHelper.getInstance();
            //执行删除操作，并返回结果
            return sql.command(cmd);
        }

        @Override
        public int changeStatus(AbsObject obj) {
            //将obj转换成目标类型
            User u=getUser(obj);
            //判断u是否为null
            if(u==null) return 0;
            //将u对象转换成用户更新的SQL指令
            String cmd="update users set status='"+u.getStatus()+
                    "' where id='"+u.getId()+"'";
            //实例化数据访问对象DBHelper
            chapter10.DBHelper sql= chapter10.DBHelper.getInstance();
            //执行用户修改操作
            return sql.command(cmd);
        }

        /*方法名：getCondition,获取查询条件
        *作用：将condition参数转换成字符串查询参数，参数condition的key必须在source中视为有效
        *@param condtion:查询条件
        *@param source:有效key
        *@return 如果转换成功，返回有效的查询字符串，否则返回null*/
        private String getCondition(Map<String,Object> condition,List<String> source){
            //判断条件的有效性
            if (condition==null || condition.size()==0 || source==null || source.size()==0) return null;
            //定义变量保存的有效条件
            StringBuilder sb=new StringBuilder();
            sb.append(" where ");
            //获取condition中的key集合
            Iterator<String> keySources=condition.keySet().iterator();
            while (keySources.hasNext()){
                //获取当前的key名称
                String key=keySources.next();
                //判断当前的key是否在source中
                if (source.contains(key))
                    if (condition.get(key) instanceof Integer)
                        sb.append( key+"="+condition.get(key).toString()+" and ");
                    else
                        sb.append( key+" like '%"+condition.get(key).toString()+"%' and ");
            }
            /*处理变量sb中的数据*/
```

```java
        if (sb.toString().equals(" where ")) return null;
        //获取最后一个and出现的位置
        int lastIndex=sb.toString().lastIndexOf("and");
        if (lastIndex>0)
            return sb.toString().substring(0,lastIndex-1).trim();
        else
            return sb.toString();
    }

    private List<String> checkColomnSource(){
        List<String> source=new ArrayList<>();
        source.add("id");
        source.add("name");
        source.add("status");
        return source;
    }
    @Override
    public List<Map<String, Object>> findResult(Map<String, Object> condition) {
        //对条件condition进行处理,转换成有效的字符串查询条件
        String queryCondition=getCondition(condition,checkColomnSource());
        //组建查询指令
        String sql="select * from users ";
        if (queryCondition!=null && !"".equals(queryCondition.trim()))
            sql+=" "+queryCondition;
        //创建数据库访问对象DBHelper
        chapter10.DBHelper dbHelper= chapter10.DBHelper.getInstance();
        //数据库执行用户查询
        return dbHelper.find(sql);
    }

    @Override
    public List<AbsObject> find(Map<String, Object> condition) {
        //查找符合条件的用户信息
        List<Map<String, Object>> result=findResult(condition);
        if (result==null || result.size()==0) return null;
        //将result转换成List<AbsObject>
        List<AbsObject> lObjs=new ArrayList<>();

        Iterator<Map<String,Object>> source=result.iterator();
        while(source.hasNext()){
            Map<String,Object> temp=source.next();
            User u=new User();
            if (temp.get("id")!=null)
                u.setId(temp.get("id").toString());
            if (temp.get("name")!=null)
                u.setName(temp.get("name").toString());
            if (temp.get("pwd")!=null)
                u.setPassword(temp.get("pwd").toString());
            if (temp.get("status")!=null)
                u.setStatus(temp.get("status").toString());
```

```java
            lObjs.add(u);
        }
        return lObjs;
    }

    @Override
    public AbsObject getMe(String id) {
        Map<String,Object> cons=new HashMap<>();
        //用户账号ID作为查询条件
        cons.put("id",id);
        //根据ID查询用户
        List<AbsObject> objs=find(cons);
        if (objs!=null && objs.size()>0 && objs.get(0) instanceof User)
            return objs.get(0);
        else
            return null;
    }

    @Override
    public int login(User user) {
        //验证参数的有效性
        if (user==null || user.getId()==null || user.getPassword()==null) return 0;
        //根据条件查询当前对象是否存在
        AbsObject temp=getMe(user.getId());
        //判断当前使用是否存在，如果不存在返回-2
        if (temp==null || temp instanceof User==false) return -2;
        //将查询结果temp转换成User用户对象
        User u=(User)temp;
        //核对密码是否一致
        if (u.getPassword().trim().equals(user.getPassword().trim()))
            return 1;//验证成功
        else
            return -3;//密码错误
    }

    @Override
    public int logout(User user) {
        return 0;
    }

    @Override
    public int changePwd(User user, String newPwd) {
        //判断用户user是否存在，如果不存在，返回null
        if (user==null || user.getId()==null) return 0;
        //判断当前用户在数据库中的数据
        AbsObject temp=getMe(user.getId());
        //不存在，返回-1
        if (temp==null) return -1;
        //将temp对象转换成User对象
        User u=(User)temp;
```

```
//将u对象转换成用户更新的SQL指令
String cmd="update users set pwd='"+newPwd+
    "' where id='"+u.getId()+"'";
//实例化数据访问对象DBHelper
chapter10.DBHelper sql= chapter10.DBHelper.getInstance();
//执行用户密码修改操作
return sql.command(cmd);
    }
}
```

2. 业务层

用户业务层由 KesunSuperService 业务层超级类、用户接口 service.IUser 和用户类 UserServiceImpl 组成。其实现代码如下：

```
package business.service.impl;
import business.dao.IDoData;
import business.dao.impl.UserDaoImpl;
import business.entity.AbsObject;
import business.entity.KesunReturn;
import business.entity.User;
import business.service.IUser;
import business.service.KesunSuperService;
/**
 * 用户管理业务层类
 * Created by wph-pc on 2018/10/31.
 */
public class UserServiceImpl extends KesunSuperService implements IUser {
    public UserServiceImpl()
    {
        //设置持久层User对象
        setModel(new User());
    }
    @Override
    public IDoData getDAO() {
        //创建User数据访问对象
        return new UserDaoImpl();
    }
    private KesunReturn checkUserAndDao(){
        KesunReturn back=new KesunReturn();
        /*返回信息初始化*/
        back.setCode("1");
        back.setMessage("初始化");
        back.setObj(1);
        if (getModel()==null || getModel() instanceof User==false){
            back.setCode("0");
            back.setMessage("系统没有获取到用户的登录信息！");
            back.setObj(null);
            return back;
        }
        IDoData dao=getDAO();
        /*判断数据访问层对象的有效性*/
        if (dao==null || dao instanceof UserDaoImpl==false){
```

```java
            back.setCode("-1");
            back.setMessage("系统没有获取到用户数据层对象!");
            back.setObj(null);
            return back;
        }
        return back;
    }
    @Override
    public KesunReturn login() {
        //检查用户参数及数据访问层参数条件是否符合
        KesunReturn back=checkUserAndDao();
        if (!back.getCode().equals("1")) return back;

        User u=(User)getModel();
        //登录
        int result=((UserDaoImpl)getDAO()).login(u);
        switch (result){
            case 1:
                back.setMessage("登录成功!");
                back.setObj(u);
                break;
            case 0:
                back.setMessage("登录用户信息错误!");
                back.setObj(null);
                break;
            case -2:
                back.setMessage("登录用户信息不存在!");
                back.setObj(null);
                break;
            case -3:
                back.setMessage("登录用户密码错误!");
                back.setObj(null);
                break;
            default:
                back.setMessage("未知错误!");
                back.setObj(null);
                break;
        }
        back.setCode(String.valueOf(result));
        return back;
    }
    @Override
    public KesunReturn logout() {
        return null;
    }
    @Override
    public KesunReturn changePwd(String newPwd) {
        //检查用户参数及数据访问层参数条件是否符合
        KesunReturn back=checkUserAndDao();
        if (!back.getCode().equals("1")) return back;
        User u=(User)getModel();
```

```
            //原密码加密
            u.setPassword(util.MyMD5.GetMyUpperMD5(u.getPassword(),u.
getId()));
            setModel(u);
            //调用原来的账号与密码进行登录
            back=login();
            if (!back.getCode().equals("1")){
                back.setCode("-1");
                back.setMessage("系统检测到当前用户原来账号信息或密码有误!");
                back.setObj(null);
                return back;
            }
            //将新密码加密
            String strPwd=util.MyMD5.GetMyUpperMD5(newPwd,u.getId());
            //登录
            int result=((UserDaoImpl)getDAO()).changePwd(u,strPwd);
            if (result>0){
                back.setMessage("密码修改成功!");
            }else
                back.setMessage("密码修改失败!");
            back.setCode(String.valueOf(result));
            back.setObj(result);
            return back;
    }
}
```

3. 表示层

用户表示层采用了 MVC 机制，它由该层的 KesunSuperController 控制层超级抽象类、用户接口 controller.IUser 和用户类 UserController 组成，其中，对 KesunSuperController 进行了相应的调整，主要是为了实现用户信息的 Ajax 数据交互。

（1）KesunSuperController 控制层超级抽象类。

KesunSuperController 控制层超级抽象类的具体实现代码如下：

```
package business.controller;
import business.entity.AbsObject;
import business.entity.KesunReturn;
import business.entity.User;
import business.service.KesunSuperService;
import com.alibaba.fastjson.JSONObject;
import util.JSONTool;

import javax.Servlet.http.HttpServlet;
import javax.Servlet.http.HttpServletRequest;
import java.io.IOException;
import java.util.Map;
import java.util.Properties;

/**
 * 控制层超级抽象类
 * Created by wph-pc on 2018/10/30.
 */
```

```java
public abstract class KesunSuperController extends HttpServlet {
    //定义业务层操作对象变量
    private KesunSuperService bll=null;
    public KesunSuperService getBll() {
        return bll;
    }
    public void setBll(KesunSuperService bll) {
        this.bll = bll;
    }
    
    //根据参数创建实体对象
    public AbsObject createModel(JSONObject jsonObject){
        //将jsonObject转换成目标类型
        return JSONTool.JSONObjectToJavaBean(jsonObject,bll.getModel().getClass());
    }
    //根据参数创建查询条件
    public abstract Map<String,Object> createCondition(JSONObject jsonObject);
    /*判断业务对象bll是否有效*/
    private KesunReturn judgeService()
    {
        KesunReturn back=new KesunReturn();
        if (bll==null){
            back.setCode("0");
            back.setMessage("业务对象为null");
            back.setObj(null);
        }
        else
            back.setObj(bll);
        return back;
    }
    /*
    * 对象新增操作
    * @param jsonObject:新增对象JSON数据
    * @return 以KesunReturn类型返回操作结果
    **/
    public KesunReturn add(JSONObject jsonObject){
        KesunReturn back=judgeService();
        if (back.getObj()==null) return back;
        //获取用户对象
        AbsObject temp=createModel(jsonObject);
        User u=(User)temp;
        //设置新增用户初始化密码,密码与账号一致
        u.setPassword(u.getId());
        bll.setModel(u);
        return bll.add();
    }
    /*
     * 对象修改操作
     * @param jsonObject:修改对象JSON数据
```

```java
 * @return 以KesunReturn类型返回操作结果
 **/
public KesunReturn edit(JSONObject jsonObject){
    KesunReturn back=judgeService();
    if (back.getObj()==null) return back;
    bll.setModel(createModel(jsonObject));
    return bll.edit();
}
/*
 * 对象删除操作
 * @param jsonObject:删除对象JSON数据
 * @return 以KesunReturn类型返回操作结果
 **/
public KesunReturn del(JSONObject jsonObject){
    KesunReturn back=judgeService();
    if (back.getObj()==null) return back;
    bll.setModel(createModel(jsonObject));
    return bll.del();
}
/*
 * 对象状态变更操作
 * @param jsonObject:变更对象JSON数据
 * @return 以KesunReturn类型返回操作结果
 **/
public KesunReturn changeStatus(JSONObject jsonObject){
    KesunReturn back=judgeService();
    if (back.getObj()==null) return back;
    bll.setModel(createModel(jsonObject));
    return bll.changeStatus();
}
/*
 * 获取单个对象操作
 * @param jsonObject:单个对象JSON数据，只需要含一个ID
 * @return 以KesunReturn类型返回操作结果
 **/
public KesunReturn getMe(JSONObject jsonObjectt){
    KesunReturn back=judgeService();
    if (back.getObj()==null) return back;
    bll.setModel(createModel(jsonObjectt));
    return bll.getMe();
}
/*
 * 以对象形式返回的数据查询操作
 * @param jsonObject:查询条件
 * @return 以KesunReturn类型返回查询结果
 **/
public KesunReturn find(JSONObject jsonObject){
    KesunReturn back=judgeService();
    if (back.getObj()==null) return back;
    Map<String,Object> cons=createCondition(jsonObject);
    return bll.find(cons);
```

```
        }
        /*
         * 以二维表格形式返回的数据查询操作
         * @param jsonObject:查询条件
         * @return 以KesunReturn类型返回查询结果
         **/
        public KesunReturn findForMap(JSONObject jsonObject){
            KesunReturn back=judgeService();
            if (back.getObj()==null) return back;
            Map<String,Object> cons=createCondition(jsonObject);
            return bll.findForMap(cons);
        }
    }
```

（2）UserController 类。

UserController 类继承于 KesunSuperController 类，且实现了该层 controller.IUser 接口，是一个 Servlet 类，该类的应用须在 web.xml 文件中配置，由于 UserController 类集成了用户管理的多个功能，而 Servlet 的映射地址只有一个，因此需要在请求页面除正常参数外，多传递一个参数 doType，由 doType 决定具体执行用户管理中的相关功能。该层用户类 UserController 在第 8 章已经设计定义，但是没有实现用户管理 CRUD 的完整功能，而且也没有采用 Ajax 标准的数据交互技术，本章将重写该类。它不仅实现了用户管理的 CRUD 功能，还实现了用户的登录及密码修改功能。具体实现代码如下：

```
package business.controller.impl;
import business.controller.IUser;
import business.controller.KesunSuperController;
import business.entity.KesunReturn;
import business.entity.User;
import business.entity.AbsObject;
import business.service.impl.UserServiceImpl;
import com.alibaba.fastjson.JSON;
import com.alibaba.fastjson.JSONObject;
import util.JSONTool;
import javax.Servlet.http.HttpServletRequest;
import javax.Servlet.http.HttpServletResponse;
import java.io.IOException;
import java.util.HashMap;
import java.util.Map;
/**
 * 用户控制层类
 * Created by wph-pc on 2018/10/31.
 */
public class UserController extends KesunSuperController implements IUser {
    public UserController()
    {
        //实例化用户业务层对象
        UserServiceImpl bll=new UserServiceImpl();
        //设置控制层的用户业务对象
        setBll(bll);
    }
```

```java
/*根据用户页面查询参数,转换成Map结构参数*/
@Override
public Map<String, Object> createCondition(JSONObject jsonObject) {
    Map<String,Object> cons=new HashMap<>();
    if (jsonObject.getString("id")!=null)
        cons.put("id",jsonObject.getString("id"));
    if (jsonObject.getString("name")!=null)
        cons.put("name",jsonObject.getString("name"));
    return cons;
}
@Override
public void doPost(HttpServletRequest request, HttpServletResponse response) throws IOException {
    KesunReturn back=new KesunReturn();
    back.setMessage("系统没有获取到任何操作!");
    back.setCode("0");
    back.setObj(null);
    /*设置编码格式及响应类型*/
    request.setCharacterEncoding("utf-8");
    response.setContentType("text/html");
    response.setCharacterEncoding("utf-8");

    //将获取的客户端请求的参数转换成JSONObject类型
    JSONObject jsonObject= JSONTool.GetRequestJSON(request);
    //获取客户端请求的操作类型
    String doType=jsonObject.getString("doType");
    switch (doType){
        case "add":
            back=add(jsonObject);
            break;
        case "edit":
            back=edit(jsonObject);
            break;
        case "del":
            back=del(jsonObject);
            break;
        case "getMe":
            back=getMe(jsonObject);
            break;
        case "login":
            back=login(jsonObject,request);
            break;
        case "loginuser":
            back=getLoginUser(request);
            break;
        case "changepwd":
            back=changePwd(jsonObject);
            break;
        case "changestatus":
            back=changeStatus(jsonObject);
            break;
```

```java
            case "find":
                back=find(jsonObject);
                break;
            case "findForMap":
                back=findForMap(jsonObject);
                break;
            default:
                back.setMessage("抱歉，系统没有获取到任何的操作指令！");
                break;
        }
        response.getWriter().write(JSON.toJSONString(back));
        response.getWriter().close();
    }
    /*
    *验证用户参数及用户业务层对象条件的有效性
    * @param jsonOjbect:客户传递的用户参数
    * @return 验证通过，返回true,否则返回false
    * */
    private Boolean checkParamCondition(JSONObject jsonObject){
        AbsObject temp=createModel(jsonObject);
        if (temp==null || temp instanceof User==false ||
                getBll()==null || getBll() instanceof UserServiceImpl==false){
            return false;
        }
        else
            return true;
    }
    @Override
    public KesunReturn login(JSONObject jsonObject,HttpServletRequest request) {
        KesunReturn back=new KesunReturn();
        //验证码验证是否存在
        if (jsonObject.getString("code")==null ||
                request.getSession().getAttribute("code")==null){
            back.setMessage("系统没有获取到登录的验证码！");
            back.setCode("0");
            back.setObj(null);
            return back;
        }
        //验证验证码的有效性
        if (!request.getSession().getAttribute("code").equals(jsonObject.getString("code"))){
            back.setMessage("验证码错误！");
            back.setCode("0");
            back.setObj(null);
            return back;
        }
        //验证参数有效性
        if (checkParamCondition(jsonObject)==false){
            back.setCode("0");
```

```java
            back.setMessage("用户或用户业务层对象为空");
            back.setObj(null);
            return back;
        }
        //获取用户参数信息
        AbsObject temp=createModel(jsonObject);
        //temp对象转换成User用户
        User u=(User) temp;
        //登录密码加密
        u.setPassword(util.MyMD5.GetMyUpperMD5(u.getPassword(),u.getId()));
        //获取用户业务层对象
        UserServiceImpl bll=(UserServiceImpl) getBll();
        //设置用户登录信息
        bll.setModel(temp);
        //登录验证
        back=bll.login();
        //如果登录成功，将登录信息写入session中
        if (back.getCode().equals("1")){
            request.getSession().setAttribute("user",temp);
        }
        //返回登录结果
        return back;
    }

    @Override
    public KesunReturn logout(JSONObject jsonObject) {
        return null;
    }
    /*实现密码修改接口*/
    @Override
    public KesunReturn changePwd(JSONObject jsonObject) {
        KesunReturn back=new KesunReturn();
        //验证参数的有效性
        if (checkParamCondition(jsonObject)==false || jsonObject.getString("newPwd")==null){
            back.setCode("0");
            back.setMessage("用户或用户业务层对象为空");
            back.setObj(null);
            return back;
        }
        //获取用户参数信息
        AbsObject temp=createModel(jsonObject);
        //获取用户业务层对象
        UserServiceImpl bll=(UserServiceImpl) getBll();
        //设置用户登录信息，包括正常的账号与密码
        bll.setModel(temp);
        //密码更改
        return bll.changePwd(jsonObject.getString("newPwd"));
```

```
    }
    /*获取当前登录的用户信息*/
    private KesunReturn getLoginUser(HttpServletRequest request){
        KesunReturn back=new KesunReturn();
        if (request.getSession().getAttribute("user")!=null){
            back.setCode("1");
            back.setMessage("已经获取到用户登录信息！");
        }else{
            back.setCode("0");
            back.setMessage("系统检测到您没有登录！");
        }
        back.setObj(request.getSession().getAttribute("user"));
        return back;
    }
}
```

开发完 UserController 类后，须在 web.xml 文件中配置 UserController 类，配置如下：
① UserController 类注册配置

```
<Servlet>
    <Servlet-name>user</Servlet-name>
    <Servlet-class>business.controller.impl.UserController</Servlet-class>
</Servlet>
```

② UserController 类地址映射配置

```
<Servlet-mapping>
    <Servlet-name>user</Servlet-name>
    <url-pattern>/user</url-pattern>
</Servlet-mapping>
```

③ doData 方法

doData 方法是 JavaScript 的方法，实现页面 Ajax 数据交互，采用了 JQuery 技术，详细定义如下：

```
/*********************************************
 * Ajax数据交互处理
 *url:数据处理请求地址
 * params:参数对象
 * callback:回调函数
 * mask:数据交互是否使用遮罩,true表示使用,不填或false表示使用
 *********************************************/
function doData(url,params,callback,mask) {
    url=getRootPath()+"/"+url;//增加项目的根地址
    $.ajax({
        type : 'post',
        url : url,
        dataType: 'json',
        contentType: "application/json; charset=utf-8",
        cache:true,
        data: JSON.stringify(params),
        beforeSend: function(){
```

```
                    if (mask!=undefined && mask==true && $("#mask").length>0)
                    {
                        $("#mask").css("height",$(document).height());
                        $("#mask").css("width",$(document).width());
                        $("#mask img").css("padding-top",window.innerHeight*0.45);
                        $("#mask").show();
                    }
                },
                complete:function () {
                    if (mask!=undefined && mask==true && $("#mask").length>0)
                        $("#mask").hide();
                },
                success: function (data) {
                    if (mask!=undefined && mask==true && $("#mask").length>0)
                        $("#mask").hide();
                    if (callback) callback(data);
                },
                error : function(arg0,arg1,arg2) {
                    switch(arg0.status)
                    {
                        case 200:
                            alert("服务器已经接收到您的请求,但无法做出正确的响应,请联系管理员进行处理,问题发生地址: " + url);
                            break;
                        case 404:
                            alert("当前操作的资源不存在,请联系管理员!");
                            break;
                        case 500:
                            alert("程序内容处理错误:500,内部符号"+url);
                            break;
                        default:
                            alert("数据处理错误,错误代码: "+arg0.status);
                            break;
                    }
                }
            });
        }
```

doData()方法中引用了getRootPath()方法,该方法返回当前项目的根地址,用户可以自己定义,参照格式:

```
/*获取网站的项目根地址*/
function getRootPath(){
    return "http://localhost:8081/jspweb";
}
```

(3)用户管理主页面设计。

用户CRUD操作集成在一个页面实现,用户CRUD页面UI设计如图10-7所示。

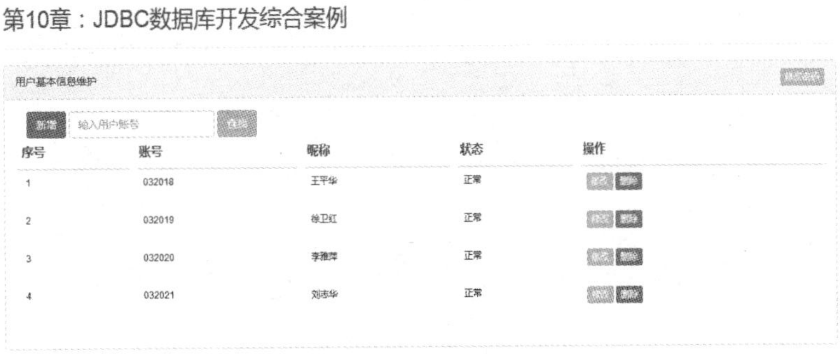

第10章：JDBC数据库开发综合案例

图 10-7　用户 CRUD 页面 UI 设计

在图 10-7 中，单击"新增"按钮，弹出如图 10-8 所示的用户新增页面。

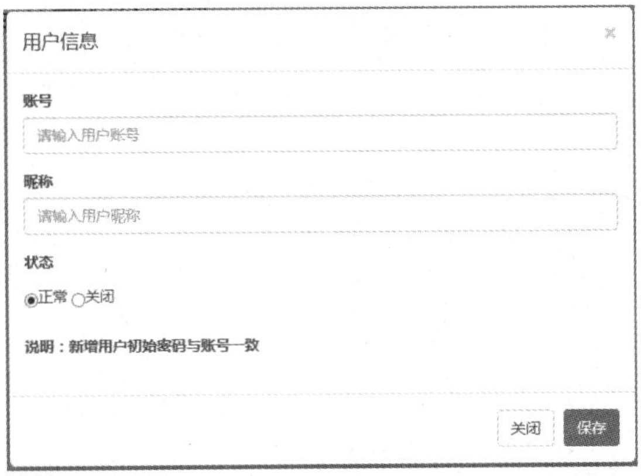

图 10-8　用户新增页面

图 10-8 中，填写完账号、昵称及状态信息，单击"保存"按钮，即可实现新用户信息的新增操作，并且新增的用户信息立即呈现在图 10-7 的表格中。在图 10-7 中，如果要删除或修改一个用户信息，建议先查找到该用户信息，在查找输入框中输入用户"账号"，支持账号模糊查询，单击"查找"按钮即可检索。如果能够查找到，查找到用户信息则呈现在数据表格中，通过单击该用户所在行最后操作列中的"修改"或"删除"按钮实现用户信息的"修改"或"删除"功能。例如查找"032018"账号，单击该行所在操作列的"修改"按钮，弹出如图 10-9 所示的用户信息修改页面

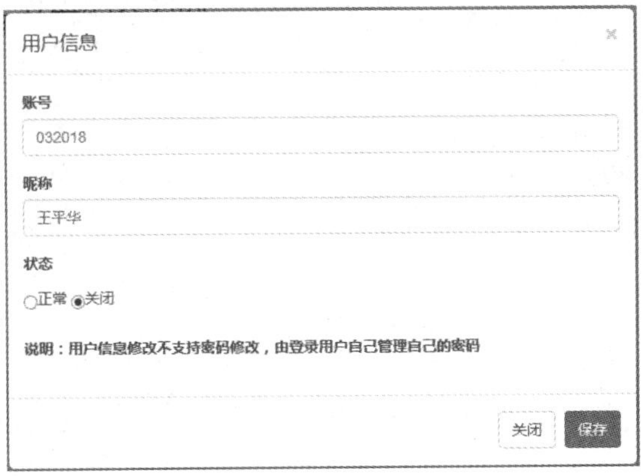

图 10-9　用户信息修改页面

如果单击用户信息所在行中的"删除"按钮，系统会弹出如图 10-10 所示的信息删除提示。

图 10-10　用户信息删除提示

单击图 10-10 中的"确定"按钮，即可执行当前用户信息的删除操作，用户数据表格信息同样会消失。

具体实现代码如下：

```jsp
<%--
  Created by IntelliJ IDEA.
  User: wph-pc
  Date: 2018/12/4
  Time: 22:03
  To change this template use File | Settings | File Templates.
--%>
<%@ page contentType="text/html;charset=UTF-8" language="java" %>
<%@ include file="../header/init_bootstrap.jsp"%>
<html>
<head>
    <title>用户管理主页面</title>
</head>
<body class="container">
<h2>第10章：JDBC数据库开发综合案例</h2>
<hr>
<div class="panel panel-default">
    <div class="panel-heading panel-primary clearfix">
        <div class="pull-left">
            用户基本信息维护
        </div >
        <div class="pull-right">
            <button class="btn btn-xs btn-warning" id="btnOpenChangeWin" data-toggle="modal">修改密码</button>
        </div>
    </div>
    <div class="panel-body">
        <div class="container-fluid form-inline">
            <button type="button" class="btn btn-primary" data-toggle="modal" id="btnAdd">新增</button>
            <input type="text" class="form-control" id="txtCondition" placeholder="输入用户账号" style="width:200px;">
            <button type="button" class="btn btn-info" id="btnFind">查找</button>
        </div>
        <div>
            <table class="table">
```

```html
                        <thead>
                            <tr><th>序号</th><th>账号</th><th>昵称</th><th>状态</th><th>操作</th></tr>
                        </thead>
                        <tbody id="tUser"></tbody>
                    </table>
                </div>
            </div>
        </div>
        <!-- 用户新增和修改对话框 -->
        <div class="modal fade" id="userModal" tabindex="-1" role="dialog" aria-labelledby="myModalLabel">
            <div class="modal-dialog" role="document">
                <div class="modal-content">
                    <div class="modal-header">
                        <button type="button" class="close" data-dismiss="modal" aria-label="Close"><span aria-hidden="true">&times;</span></button>
                        <h4 class="modal-title" id="myModalLabel">用户信息</h4>
                    </div>
                    <div class="modal-body">
                        <form id="ff">
                            <div class="form-group">
                                <label for="txtID">账号</label>
                                <input type="text" class="form-control" id="txtID" name="id" placeholder="请输入用户账号">
                            </div>
                            <div class="form-group">
                                <label for="txtName">昵称</label>
                                <input type="text" class="form-control" id="txtName" name="name" placeholder="请输入用户昵称">
                            </div>
                            <div class="form-group form-inline">
                                <label for="status">状态</label>
                                <div id="status">
                                    <input type="radio" class="form-control" name="status" value="正常" checked="checked">正常
                                    <input type="radio" class="form-control" name="status" value="关闭">关闭
                                </div>
                            </div>
                            <div class="form-group">
                                <label id="lblTips"></label>
                            </div>
                        </form>
                    </div>
                    <div class="modal-footer">
                        <button type="button" class="btn btn-default" data-dismiss="modal">关闭</button>
                        <button type="button" class="btn btn-primary" id="btnSave">保存</button>
                    </div>
```

```html
                </div>
            </div>
        </div>
        <!--用户密码修改对话框-->
        <div class="modal fade" id="changePwdModal" tabindex="-2" role="dialog" aria-labelledby="pwdModalLabel">
            <div class="modal-dialog" role="document">
                <div class="modal-content">
                    <div class="modal-header">
                        <button type="button" class="close" data-dismiss="modal" aria-label="Close"><span aria-hidden="true">&times;</span></button>
                        <h4 class="modal-title" id="pwdModalLabel">密码修改</h4>
                    </div>
                    <div class="modal-body">
                        <form id="frmChangePwd">
                            <div class="form-group">
                                <label for="txtAccountID">账号</label>
                                <input type="text" readonly="readonly" class="form-control" id="txtAccountID" name="id" placeholder="请输入用户账号">
                            </div>
                            <div class="form-group">
                                <label for="txtOldPwd">原密码</label>
                                <input type="password" class="form-control" id="txtOldPwd" name="password" placeholder="请输入原密码">
                            </div>
                            <hr>
                            <div class="form-group">
                                <label for="txtNewPwd">新密码</label>
                                <input type="password" class="form-control" id="txtNewPwd" name="password" placeholder="请输入新密码">
                            </div>
                            <div class="form-group">
                                <label for="txtNewPwd2">新密码确认</label>
                                <input type="password" class="form-control" id="txtNewPwd2" name="password" placeholder="请再次输入新密码,并且保证两次的输入一致!">
                            </div>
                        </form>
                    </div>
                    <div class="modal-footer">
                        <button type="button" class="btn btn-default" data-dismiss="modal">关闭</button>
                        <button type="button" class="btn btn-primary" id="btnChangePwd">变更密码</button>
                    </div>
                </div>
            </div>
        </div>

        <script>
            function find() {
```

```javascript
            $("#tUser").empty();
            var con=new Object();
            con.doType="findForMap";
            con.id=$("#txtCondition").val();

            doData("user",con,function (result) {
                if (result.obj!=undefined && result.obj.length>0)
                    for(var i=0;i<result.obj.length;i++)
                        $("#tUser").append("<tr><td>"+(i+1)+"</td><td>"+result.obj[i].id+
                            "</td><td>"+result.obj[i].name+"</td><td>"+result.obj[i].status+
                            "</td><td><button type='button' class='btn btn-xs btn-warning' action='edit' objId='"+
                            result.obj[i].id+"' data-toggle='modal'>修改</button> <button type='button' class='btn btn-xs btn-danger' action='del' objId='"+
                            result.obj[i].id+"' objName='"+result.obj[i].name+"'>删除</button></td></tr>");
            },true);
        }
        //是否新增操作，true表示新增，false表示修改
        var isAdd=true;
        $(function () {
            //查询数据
            find();
            $("#btnAdd").click(function () {
                isAdd=true;
                $("#lblTips").text("说明：新增用户初始密码与账号一致");
                $("#txtID").val("");
                $("#txtName").val("");
                $("#userModal").modal("show");
            });
            $("#btnFind").click(function () {
                find();
            });
            $(document).on("click","button[action=edit]",function () {
                var id=$(this).attr("objId");
                /*获取当前账号信息*/
                var user=new Object();
                user.id=id;
                user.doType="getMe";
                doData("user",user,function (data) {
                    if (data.obj!=undefined && data.obj!=null){
                        $("#txtID").val(data.obj.id);
                        $("#txtName").val(data.obj.name);
                        setRadioChecked("status",data.obj.status);
                        isAdd=false;
                        $("#lblTips").text("说明：用户信息修改不支持密码修改，由登录用户自己管理自己的密码");
                        $("#userModal").modal("show");
```

```javascript
            }else
                toastr.info(data.message);
        },true);
    });
    $(document).on("click","button[action=del]",function () {
        var id=$(this).attr("objId");
        var flag=confirm("您正在执行删除用户【"+$(this).attr("objName")+"】,删除后不可恢复,确定吗? ");
        if(flag){
            var user=new Object();
            user.id=id;
            //设置操作类型
            user.doType="del";
            doData("user",user,function (data) {
                if (data.code>0){
                    find();//刷新数据
                }
                //提示操作结果
                toastr.info(data.message);
            },true);
        }
    });
    $("#btnSave").click(function () {
        var user=serializeArrayToObject("ff");
        if (isAdd)
            user.doType="add";
        else
            user.doType="edit";
        doData("user",user,function (data) {
            if (data.code>0){
                find();
                $("#userModal").modal("hide");
            }
            toastr.info(data.message);
        },true);
    });

    $("#btnOpenChangeWin").click(function () {
        /*获取当前登录的用户信息*/
        var obj=new Object();
        obj.doType="loginuser";
        doData("user",obj,function (data) {
            if (data.code=="0")
                toastr.info(data.message);
            else{
                $("#txtAccountID").val(data.obj.id);
                $("#changePwdModal").modal("show");
            }
        },true);
    });
    $("#btnChangePwd").click(function () {
```

```javascript
            if ($("#txtNewPwd").val()=="" || $("#txtOldPwd").val()==""){
                toastr.info("原密码和新密码都不能为空！");
                return;
            }
            if ($("#txtNewPwd").val()!=$("#txtNewPwd2").val()){
                toastr.info("新密码两次都不一致！");
                return;
            }
            var obj=new Object();
            obj.id=$("#txtAccountID").val();
            obj.doType="changepwd";
            obj.password=$("#txtOldPwd").val();
            obj.newPwd=$("#txtNewPwd").val();
            doData("user",obj,function (data) {
                if (data.code>0)
                    location.href="userLogin.jsp";
                toastr.info(data.message);
            },true);
        });
    });
</script>
</body>
</html>
```

上面代码中出现的 toastr 对象是 BootStrap 提示信息的一个插件，doData()方法是 public.js 中封装的一个方法。这些代码不仅实现了页面构建，而且已经将用户 CRUD 功能通过脚本 JavaScript 技术与 UserController 控制层进行整合，实现了用户管理的相应功能。

10.3 登录与 MD5 密码管理

登录与密码管理不可分离，用户登录时，需要提供登录的秘钥凭证，密码管理则非常重要，如果将用户密码直接放在数据中，就会很容易泄密。因此，数据库中存放用户的密码应该是加密后的数据，即使是程序员看到数据库的密码，也不知道具体用户密码的明文。

案例描述

在用户 CRUD 管理的基础之上，实现用户登录，具体要求：
- 实现用户登录时的身份验证，要求与数据库进行真实比对；
- 实现用户 MD5 密码加密，并且要求实现即使不同用户采用相同的密码，加密后的数据库密码密文也不一样；
- 实现用户密码修改，操作者只能修改自己的密码。

案例分析

用户登录与密码安全管理是信息系统安全防范的重要手段之一，特别是用户密码安全机制。根据案例描述，对于用户密码管理准备采用 MD5 加密手段，再给每个加密的密码另外增加一个"盐值"来控制不同用户采用相同的密码时，加密后的数据库密码密文也不一样，这个"盐值"是使用用户的账号，因为账号具有唯一性，这样，每个用户的真正密码就是"密码+账号"，再通过 MD5 进行加密即可。

为了更好地管理用户密码加密问题，本章设计了一个 MyMD5 类，该类的设计见表 10-8。

表 10-8 MyMD5 类

类 名 称	MyMD5	中文名称	加密类	类 型	实体类	父类	无
类 描 述	实现 MD5 数据加密功能							
方　法							
序 号	方法名	参　数		返回类型	功能说明		
1	GetMyLowerMD5	String old：加密数据，String salt：加密"盐值"		string	类方法，实现对字符串 old 加密，并使用 salt 盐值，返回加密后的数据，数据使用十六进制，字母部分采用小写		
2	GetMyUpperMD5	String old：加密数据，String salt：加密"盐值"		string	类方法，实现对字符串 old 加密，并使用 salt 盐值，返回加密后的数据，数据使用十六进制，字母部分采用大写		

MyMD5 类的代码实现部分：

```java
package util;

import java.math.BigInteger;
import java.security.MessageDigest;

/**
 * MD5加密
 * Created by wph-pc on 2018/12/8.
 */
public class MyMD5 {
    /*
     * 对字符串old进行MD5加密
     * @param old:加密字符串
     * @param salt:加密盐值
     * @return 返回加密后的字符串,字母为小写形式
     **/
    public static String GetMyLowerMD5(String old,String salt) {
        try {
            // 创建一个MD5加密计算对象
            MessageDigest md = MessageDigest.getInstance("MD5");
            // 加密计算
            md.update((old+salt).getBytes());
            // 转换成有效的hash值
            return new BigInteger(1, md.digest()).toString(16);
        } catch (Exception e) {
            e.printStackTrace();
            return null;
        }
    }
    /*
```

```
 * 对字符串old进行MD5加密
 * @param old:加密字符串
 * @param salt:加密盐值
 * @return 返回加密后的字符串,字母为大写形式
 */
public static String GetMyUpperMD5(String old,String salt) {
    char hexDigits[]={'0','1','2','3','4','5','6','7','8','9','A','B','C','D','E','F'};
    try {
        //将加密的字符串与盐值组合成新的加密字符串,并转换成数组
        byte[] bTemp = (old+salt).getBytes();
        // 获得MD5的 MessageDigest 对象
        MessageDigest md = MessageDigest.getInstance("MD5");
        // 使用需要加密字节数组更新
        md.update(bTemp);
        // 获得密文
        byte[] bAfter = md.digest();
        // 把密文转换成十六进制的字符串形式
        int len = bAfter.length;
        char str[] = new char[len * 2];
        int k = 0;
        for (int i = 0; i < len; i++) {
            byte byte0 = bAfter[i];
            str[k++] = hexDigits[byte0 >>> 4 & 0xf];
            str[k++] = hexDigits[byte0 & 0xf];
        }
        return new String(str);
    } catch (Exception e) {
        e.printStackTrace();
        return null;
    }
}
```

说明：用户密码初始化在新增的时候，设置用户密码与账号一致。

案例实现

用户登录与密码修改的相关功能实际上在用户CRUD的案例模块中都已经给出了详细的实现代码，补充说明如下：

1. 数据访问层

- 数据访问层用户登录功能，请参照 UserDaoImpl 类中的 login()方法。
- 数据访问层用户密码修改功能，请参照 UserDaoImpl 类中的 changePwd()方法。

2. 业务层

- 业务层用户登录功能，请参照 UserServiceImpl 类中的 login()方法。
- 业务层用户密码修改功能，请参照 UserServiceImpl 类中的 changePwd()方法，加密关键代码：

```
//将新密码newPwd加密
String strPwd=util.MyMD5.GetMyUpperMD5(newPwd,u.getId());
```

3. 表示层

- 表示层用户登录功能，请参照 UserController 类中的 login()方法。
- 表示层用户密码修改功能，请参照 UserController 类中的 changePwd()方法。

4. 登录页面实现

登录页面采用 BootStrap 实现，需要输入用户的账号、密码及验证码，如图 10-11 所示。

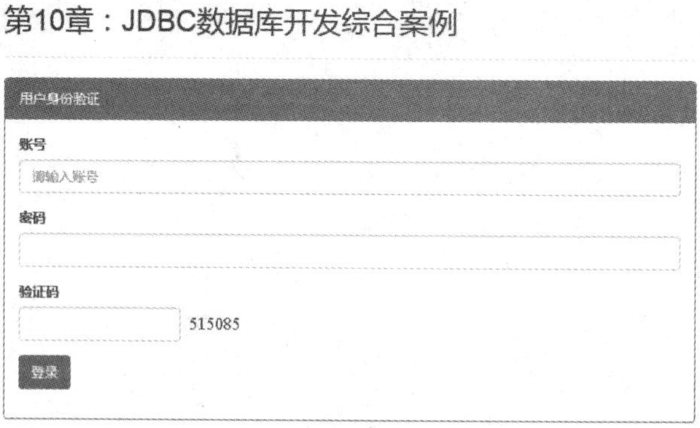

图 10-11　用户登录

用户登录页面 userLogin.jsp 中，验证码采用了 response 内置对象章节中的案例，具体的页面实现及 JavaScript 功能实现如下：

```jsp
<%--
  Created by IntelliJ IDEA.
  User: wph-pc
  Date: 2018/12/8
  Time: 14:57
  To change this template use File | Settings | File Templates.
--%>
<%@ page contentType="text/html;charset=UTF-8" language="java" %>
<%@include file="../header/init_bootstrap.jsp"%>
<html>
<head>
    <title>用户登录【数据库章节】</title>
</head>
<body class="container">
<h2>第10章：JDBC数据库开发综合案例</h2>
<hr>

<div class="panel panel-primary">
    <div class="panel-heading panel-primary">
        用户身份验证
    </div>
    <div class="panel-body">
        <div class="row">
            <div class="col-md-12 col-sm-12 col-lg-12 col-xs-12">
                <form method="post">
                    <div class="form-group">
                        <label for="txtAccount">账号</label>
                        <input type="text" class="form-control" name="id"
```

```html
placeholder="请输入账号" id="txtAccount"/>
                    </div>
                    <div class="form-group">
                        <label for="txtPwd">密码</label>
                        <input type="password" class="form-control" name="password" id="txtPwd"/>
                    </div>
                    <div class="form-group">
                        <label for="txtCode">验证码</label>
                        <div class="form-inline">
                            <input type="text" class="form-control" name="code" maxlength="6" id="txtCode"/>
                            <img src="../chapter6/response/codeimage.jsp" width="60" height="20" id="imgCode" onclick="refreshImg()">
                        </div>
                    </div>
                    <button type="button" class="btn btn-primary" id="btnLogin">登录</button>
                </form>
            </div>
        </div>
    </div>
</div>


<script>
    function refreshImg() {
        document.getElementById("imgCode").src="../chapter6/response/codeimage.jsp?timestamp="+(new Date()).getTime();
    }
    $(function () {

        $("#btnLogin").click(function () {
            var user=new Object();
            user.id=$("#txtAccount").val();
            user.password=$("#txtPwd").val();
            user.code=$("#txtCode").val();
            //操作类型参数
            user.doType="login";
            doData("user",user,function (data) {
              toastr.info(data.message);
              if (data.code>0)
                  location.href="UserManage.jsp";
              //验证码刷新
              refreshImg();
            });
        });
    });
</script>
</body>
</html>
```

5. 密码修改页面实现

密码修改页面在用户综合管理页面中，单击图 10-7 中的"密码修改"按钮，弹出的密码修改页面如图 10-12 所示。

图 10-12 密码修改页面

图 10-12 中，修改密码页面中的用户是当前登录者，只能自己修改自己的密码，在修改密码时，需要提供正确的原密码和两次一致的新密码，系统先验证原密码是否正确，如果错误，将不会启用新密码，如果验证原密码无误，且两次新密码一样，则修改成功，启用新密码重新登录。

运行结果

启动运行用户登录页面 userLogin.jsp，输入正确的账号、密码及验证码，就可以进入用户管理主页面。

习题

在第 8 章实现对权限基本信息的维护中，数据访问层的功能没有实现，本项目的习题请继续完善第 8 章的项目，具体要求如下：

1. 按照权限基本信息属性，采用 MySQL 数据库，建立权限基本信息表；
2. 采用 JDBC 技术，利用连接池建立 Java 与 MySQL 数据库连接；
3. 数据库的访问采用 DBHelper 实现；
4. 实现权限基本信息维护中的数据访问层规定的各项功能，包括权限基本信息新增、修改、删除及查找功能；
5. 通过页面测试权限基本信息维护的各项功能。

第 11 章 文件上传与下载

文件上传与下载技术在系统开发中非常重要,例如:将上传本人的照片、提交作业附件等都属于文件上传的范畴;同样,文件下载技术的应用也非常广泛,从 Web 应用服务器上下载资源。本章的文件上传与下载技术属于两种不同的技术,文件上传是指 Web 客户端将文件(可以是任何形式的文件)上传到 Web 服务器上。文件下载是指客户端从服务器上安全地获取指定的文件资源。

本章任务

(1)文件上传与下载相关技术;
(2)文件上传技术的应用;
(3)文件安全下载技术应用。

重点内容

(1)掌握文件上传并与数据库同步;
(2)掌握文件安全下载技术。

难点内容

(1)文件上传并与数据库同步;
(2)文件安全下载技术。

11.1 文件上传与下载相关知识

文件上传与下载在项目中的应用非常广泛,特别是 Web 应用程序,为了集中管理文件资源,需要将客户端的文件上传到服务器指定地方存放,需要的时候再通过下载技术将资源从服务器端下载到客户端。JSP 中的文件上传必须通过 apache 提供的用户文件读写操作

commons-io 和文件上传 commons-fileupload 的两个 jar 包实现。文件下载本章通过 IO 流读写方式实现,而不是简单的链接方式。

11.1.1 文件上传相关知识

JSP 中的文件上传使用了 commons-fileupload 中的 FileItem 接口、FileItemFactory 接口、DiskFileItemFactory 类和 ServletFileUpload 类等实现。

1. FileItem 接口

FileItem 接口封装单个表单字段元素的数据,一个表单字段元素对应一个 FileItem 对象,通过调用 FileItem 对象的方法可以获得相关表单字段元素的数据。FileItem 是一个接口,在应用程序中使用的实际上是该接口的一个实现类,该实现类的名称并不重要,程序可以采用 FileItem 接口类型来对它进行引用和访问。该接口的描述见表 11-1。

表 11-1 FileItem 接口描述

类名称	FileItem	中文名称	上传文件项	类型	接口
类描述	FileItem 接口封装单个表单字段元素的数据,一个表单字段元素对应一个 FileItem 对象,通过调用 FileItem 对象的方法可以获得相关表单字段元素的数据				
序号	方法名	参数	返回类型	功能说明	
1	isFormField	无	boolean	判断 FileItem 类对象封装的数据是属于一个普通表单字段,还是属于一个文件表单字段,如果是普通表单字段则返回 true,否则返回 false	
2	getName	无	string	getName 方法用于获得文件上传字段中的文件名,如果 FileItem 类对象对应的是普通表单字段,getName 方法将返回 null	
3	getFieldName	无	string	获取表单字段元素描述头的 name 属性值,也是表单标签 name 属性的值	
4	write	File file: 写入的目标文件	void	将 FileItem 对象中保存的主体内容保存到某个指定的文件中。如果 FileItem 对象中的主体内容是保存在某个临时文件中,该方法顺利完成后,临时文件有可能会被清除	
5	getString	无	string	将 FileItem 对象中保存的主体内容作为一个字符串返回	
6	getString	String encoding: 字符编码	string	将 FileItem 对象中保存的主体内容作为一个字符串返回	
7	getContentType	无	string	用于获取文件上传的类型	
8	isInMemory	无	boolean	用来判断 FileItem 类对象封装的主体内容是存储在内存中,还是存储在临时文件中,如果存储在内存中则返回 true,否则返回 false	
9	delete	无	void	用来清空 FileItem 类对象中存放的主体内容,如果主体内容被保存在临时文件中,delete 方法将删除该临时文件	

2. DiskFileItemFactory 类

将请求消息实体中的每个项目封装成单独的 DiskFileItem（FileItem 接口的实现）对象。org.apache.commons.fileupload.FileItemFactory 接口的实现由 org.apache.commons.fileupload.disk.DiskFileItemFactory 来完成。当上传的文件项目比较小时，直接保存在内存中（速度比较快）；文件项目比较大时，以临时文件的形式保存在磁盘临时文件夹（虽然速度慢些，但是内存资源是有限的）。该类的属性包含了三个属性：

- public static final int DEFAULT_SIZE_THRESHOLD：将文件保存在内存还是磁盘临时文件夹的默认临界值，值为 10240 字节。
- private File repository：用于配置在创建文件项目时，当文件项目大于临界值时使用的临时文件夹，默认采用系统临时文件路径，可以通过系统属性 java.io.tmpdir 获取。
- private int sizeThreshold：将文件保存在内存还是磁盘临时文件夹的临界值。

DiskFileItemFactory 类的设计见表 11-2。

表 11-2　DiskFileItemFactory 类的设计

类名称	DiskFileItemFactory	中文名称	磁盘文件项工厂	类型	实体类
类描述	提供文件项 FileItem 磁盘操作，实现接口 FileItemFactory					
序号	方法名	参数	返回类型	功能说明	
1	DiskFileItemFactory	无	构造方法	构造方法，采用默认临界值和系统临时文件夹构造文件项工厂对象	
2	DiskFileItemFactory	int sizeThreshold：临界值；File repository：临时文件	构造方法	构造方法，采用参数指定临界值和系统临时文件夹构造文件项工厂对象	
3	createItem	无	FileItem	根据 DiskFileItemFactory 相关配置将每个请求消息实体项目创建成 DiskFileItem 实例，并返回。该方法无须调用，FileUpload 组件在解析请求时内部使用	
4	setSizeThreshold	int sizeThreshold：临界值	void	设置是否将上传文件以临时文件的形式保存在磁盘的临界值（以字节为单位的 int 值）	
5	setSizeThreshold	无	void	采用系统默认值 10KB 作为临界值。对应的 getSizeThreshold() 方法用来获取此临界值	
6	setRepository	File repository：临时文件目录	void	用于设置当上传文件尺寸大于 setSizeThreshold 方法设置的临界值时，文件将以临时文件形式保存在磁盘上的存放目录	

3. ServletFileUpload 类

ServletFileUpload 类是 Apache 组件处理文件上传的核心高级类，通过使用 parseRequest

(HttpServletRequest)方法可以将 HTML 中每个表单提交的数据封装成一个 FileItem 对象，以 List 列表的形式返回。ServletFileUpload 类的设计见表 11-3。

表 11-3 ServletFileUpload 类的设计

类 名 称	ServletFileUpload	中文名称	文件上传类	类　　型	实体类
类 描 述	提供文件上传功能，将表单提交的数据转换成 FileItem 对象，并以 List 返回					
序　　号	方 法 名	参　　数	返回类型	功能说明	
1	ServletFileUpload	无	构造方法	构造一个未初始化的实例，需要在解析请求之前先调用 setFileItemFactory() 方法设置 fileItemFactory 属性	
2	ServletFileUpload	FileItemFactory fileItemFactory：文件项工厂	构造方法	构造一个实例，并根据参数指定的 FileItemFactory 对象，设置 fileItemFactory 属性	
3	setSizeMax	long sizeMax：上传文件最大值	void	用于设置请求消息实体内容（即所有上传数据）的最大尺寸限制，以防止客户端恶意上传超大文件来浪费服务器端的存储空间。其参数是以字节为单位的 long 型数字	
4	setFileSizeMax	long fileSizeMax：单个文件上传最大值	void	用于设置单个上传文件的最大尺寸限制，以防止客户端恶意上传超大文件来浪费服务器端的存储空间。其参数是以字节为单位的 long 型数字	
5	parseRequest	HttpServletRequest req：请求内置对象	List	该方法是 ServletFileUpload 类的重要方法，它是对 HTTP 请求消息体内容进行解析的入口方法。它解析出 FORM 表单中的每个字段的数据，并将它们分别包装成独立的 FileItem 对象，然后将这些 FileItem 对象加入一个 List 类型的集合对象中返回	
6	getItemIterator	HttpServletRequest request：请求内置对象	FileItemIterator	该方法和 parseRequest 方法基本相同。但是 getItemIterator 方法返回的是一个迭代器，该迭代器中保存的不是 FileItem 对象，而是 FileItemStream 对象，性能较好	
7	isMultipartContent	HttpServletRequest req：请求内置对象	boolean	静态方法，用于判断请求消息中的内容是否是"multipart/form-data"类型，是则返回 true，否则返回 false	
8	setProgressListener	ProgressListener pListener：上传进度监听器对象	void	设置文件上传进度监听器	

11.1.2　文件下载相关知识

文件下载在项目实际应用中比较广泛，例如从网站上下载图片、歌曲、附件等都需要利用文件下载技术。本章讲述的文件下载技术是采用 IO 流的方式，从服务器上读取下载目标文件的内容，写入 response 内置对象提供的输出流对象中，从而实现文件下载，这样隐藏了下载文件存放的位置。不仅如此，本节设计的下载技术还要数据库的配合，页面上提供的下载文件 ID 并非真正的文件名，需要从数据库中查询才能得到真正的文件名，这种技术下载比较安全。实现文件下载，需要使用 IO 流技术，主要涉及字节输入流、字节输出流及缓冲流技术。

1. InputSream 字节输入流

InputStream 字节输入流实现对非字符型数据的读取操作，例如：图片、Word、Excel、视频、音频等各种文件。InputStream 字节输入流是抽象类，具体功能的应用需要通过其子类调用；字节输入流父子关系如图 11-1 所示。

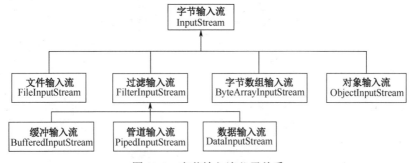

图 11-1　字节输入流父子关系

从图 11-1 中可以看出，InputStream 字节输入流的子类包括文件输入流 FileInputStream、过滤输入流 FilterInputStream、字节数组输入流 ByteArrayInputStream 和对象输入流 ObjectInputStream；其中，过滤输入流有管道输入流 PipedInputStream、数据输入流 DataInputStream 和缓冲输入流 BufferedInputStream 3 个子类。

InputStream 字节输入流提供了字节数据读取的操作功能，默认缓冲 2048 字节，InputStream 类描述参照表 11-4。

表 11-4　InputStream 类描述

类 名 称	InputStream	中文名称	字节输入流	类　　型	抽象类
类 描 述	提供了字节数据读取的操作功能				
序　号	方 法 名	参　　数		返回类型	功能说明
1	read	无		int	抽象方法，子类实现，从输入流中读取一个字节，整数范围，值范围 0~255
2	read	byte b[]： 存放读取的字节数据。		int	将读取的字节数据存放在参数 b 中，返回读取的字节个数
3	read	byte b[]： 存放读取的字节数据； int off：开始读取位置； int len：读取的长度		int	将读取的字节数据存放在参数 b 中从位置 off 开始，长度为 len，返回读取的字节个数
4	close	无		void	读取完数据后，流关闭

（1）FileInputStream 文件输入流。

FileInputStream 文件输入流实现对文件进行读取操作。该类的描述见表 11-5。

表 11-5 FileInputStream 类描述

类 名 称	FileInputStream	中文名称	文件输入流	类 型	实体类
类 描 述	实现对文件进行读取操作				
序 号	方法名	参 数	返回类型	功能说明	
1	FileInputStream	String name：读取文件名	FileInputStream	根据参数 name 创建文件输入流对象	
2	FileInputStream	File file：读取文件对象	FileInputStream	根据参数 file 创建文件输入流对象	

（2）FilterInputStream 过滤输入流。

FilterInputStream 过滤输入流提供了用来封装其他的输入流，并为它们提供额外的功能；该类常用的子类有 BufferedInputStream 缓冲输入流类、PipedInputStream 管道输入字节流和 DataInputStream 数据输入字节流 3 个子类。

BufferedInputStream 缓冲输入流

BufferedInputStream 缓冲输入流为其他字节输入流的读取提供了缓冲读取操作，实体类。该类的描述见表 11-6。

表 11-6 BufferedInputStream 类描述

类 名 称	BufferedInputStream	中文名称	缓冲输入流	类 型	实体类
类 描 述	为其他字节输入流的读取提供了缓冲读取操作				
序 号	方法名	参 数	返回类型	功能说明	
1	BufferedInputStream	InputStream in：输入字节流对象	BufferedInputStream	根据参数 in 创建缓冲字节输入流对象	
2	BufferedInputStream	InputStream in：输入字节流对象；int size：缓冲大小	BufferedInputStream	根据参数 in 和缓冲参数 size 创建缓冲字节输入流对象	

PipedInputStream 管道输入流

PipedInputStream 管道输入流提供了多线程之间进行通信字节数据的读取操作。该类的描述见表 11-7。

表 11-7 PipedInputStream 类描述

类 名 称	PipedInputStream	中文名称	管道输入流	类 型	实体类
类 描 述	提供了多线程之间进行通信字节数据的读取操作				
序 号	方法名	参 数	返回类型	功能说明	
1	PipedInputStream	PipedOutputStream src：管道输出流对象	管道输入流对象	构造方法，根据参数 src 管道输出流对象构建 PipedInputStream 对象	
2	PipedInputStream	PipedOutputStream src：管道输出流对象；int pipeSize：管道缓冲大小	管道输入流对象	构造方法，根据参数 src 管道输出流对象和 pipeSize 缓冲大小构建 PipedInputStream 对象	

续表

序号	方法名	参数	返回类型	功能说明
3	PipedInputStream	int pipeSize：管道缓冲大小	管道输入流对象	构造方法，根据参数 pipeSize 缓冲大小构建 PipedInputStream 对象
4	PipedInputStream	无	管道输入流对象	构造方法，根据默认缓冲大小构建 PipedInputStream 对象
5	connect	PipedOutputStream src：管道输出流对象	void	连接到参数 src 输出流管道

DataInputStream 数据输入流

DataInputStream 数据输入流用来专门读取各种基本数据类型的数据，需要与 DataOutputStream 数据输出流配合使用。DataInputStream 数据输入流类的描述见表 11-8。

表 11-8　DataInputStream 数据输入流类的描述

类名称	DataInputStream	中文名称	数据输入流	类型	实体类
类描述	读取各种基本数据类型的数据				
序号	方法名	参数	返回类型	功能说明	
1	DataInputStream	InputStream src：字节输入流对象	数据输入流对象	构造方法，根据参数 src 字节输入流对象构建 DataInputStream 对象	
2	readFully	byte b[]：数据读取存放处	void	从数据源中读取数据并存放在参数 b 中	
3	readFully	byte b[]：数据读取存放处；int off：读取数据位置；int len：读取数据的长度	void	从数据源中读取数据，从参数 off 位置开始，读取长度为 len 并存放在参数 b 中	
4	readBoolean	无	boolean	从数据源中读取布尔型数据。	
5	readByte	无	byte	从数据源中读取一个有符号字节数据	
6	readUnsignedByte	无	byte	从数据源中读取一个无符号字节数据	
7	readShort	无	short	从数据源中读取一个有符号短整型数据	
8	readUnsignedShort	无	short	从数据源中读取一个无符号短整型数据	
9	readChar	无	char	从数据源中读取一个字符型数据	
10	readInt	无	int	从数据源中读取一个整型数据	
11	readLong	无	long	从数据源中读取一个长整型数据	
12	readFloat	无	float	从数据源中读取一个单精度浮点型数据	
13	readDouble	无	double	从数据源中读取一个双精度浮点型数据	

（3）ByteArrayInputStream 字节数组输入流。

ByteArrayInputStream 字节数组输入流实现将数据源中的数据读取到字节数组中，该类的描述见表 11-9。

表 11-9　ByteArrayInputStream 类描述

类 名 称	ByteArrayInputStream	中文名称	字节数组输入流	类　型	实体类
类 描 述	将数据源中的数据读取到字节数组中				
序　号	方 法 名	参　　数	返回类型	功能说明	
1	ByteArrayInputStream	byte buf[]： 存放读入的字节内容	字节数组输入流对象	构造方法，根据参数 buf 对象，构建 ByteArrayInputStream 对象	
2	ByteArrayInputStream	byte buf[]： 存放读入的字节内容； int offset：读取的位置； int length：读取的长度	字节数组输入流对象	构造方法，根据参数 buf 对象，读取位置 offset 和读取长度 length 构建 ByteArrayInputStream 对象	

（4）ObjectInputStream 对象输入流。

ObjectInputStream 对象输入流实现提取存放在对象中的文件，也可以读取对象中的 Java 基本数据类型数据。该类的描述见表 11-10。

表 11-10　ObjectInputStream 类描述

类 名 称	ObjectInputStream	中文名称	对象输入流	类　型	实体类
类 描 述	实现提取存放在对象中的文件				
序　号	方 法 名	参　　数	返回类型	功能说明	
1	ObjectInputStream	InputStream in： 输入流对象	对象输入流对象	构造方法，根据参数 in 对象构建 ObjectInputStream 对象	
2	readObject	无	Object	读取当前输入流中的对象信息，并返回对象	
3	readFields	无	GetField	读取对象中所有的属性名称	
4	readFully	byte b[]： 数据读取存放处	void	从读取对象中读取数据并存放在参数 b 中	
5	readFully	byte b[]：数据读取存放处； int off：读取数据位置； int len：读取数据的长度	void	从读取对象中读取数据，从参数 off 位置开始，读取长度为 len 并存放在参数 b 中	
6	readBoolean	无	boolean	从读取对象中读取布尔型数据	
7	readByte	无	byte	从读取对象中读取一个有符号字节数据	
8	readUnsignedByte	无	byte	从读取对象中读取一个无符号字节数据	
9	readShort	无	short	从读取对象中读取一个有符号短整型数据	
10	readUnsignedShort	无	short	从读取对象中读取一个无符号短整型数据	
11	readChar	无	char	从读取对象中读取一个字符型数据	
12	readInt	无	int	从读取对象中读取一个整型数据	
13	readLong	无	long	从读取对象中读取一个长整型数据	

2. OutputStream 字节输出流

OutputStream 字节输出流提供了对字节流进行写入（输出）操作，它是一个抽象类，需要通过子类实现其提供的相应功能。如图 11-2 所示为 OutputStream 类与子类之间的关系。

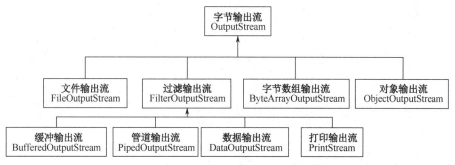

图 11-2　OutputStream 类与子类之间的关系

OutputStream 字节输出流下面有 FileOutputStream 文件输出流、FilterOutputStream 过滤输出流、ByteArrayOutputStream 字节数组输出流和 ObjectOutputStream 对象输出流 4 个子类；其中，FilterOutputStream 过滤输出流下有 BufferedOutputStream 缓冲输出流、PipedOutputStream 管道输出流、DataOutputStream 数据输出流和 PrintStream 打印输出流 4 个子类。OutputStream 类的描述见表 11-11。

表 11-11　OutputStream 类描述

类 名 称	OutputStream	中文名称	字节输出流	类 型	抽象类
类 描 述	提供对字节流进行写入（输出）操作				
序 号	方 法 名	参　　数	返回类型	功能说明	
1	write	int b：写入的字节	void	抽象方法，将字节参数 b 写入流中	
2	write	byte[] b：写入的字节数组	void	将字节数组参数 b 写入流中	
3	write	byte[] b：写入的字节数组； int off：写入起始位置； int len：写入的长度	void	将字节数组参数 b，从 off 位置开始，长度为 len 的字节数据写入流中	
4	flush	无	void	将缓存中的数据全部写入流中，并清空缓存	
5	close	无	void	关闭输出流	

（1）FileOutputStream 文件输出流。

FileOutputStream 文件输出流实现对字节文件进行写入操作，例如：图片文件、视频文件、音频文件等。FileOutputStream 类描述见表 11-12。

表 11-12　FileOutputStream 类描述

类 名 称	FileOutputStream	中文名称	文件输出流	类 型	实体类
类 描 述	实现对字节文件进行写入操作				
序 号	方 法 名	参　　数	返回类型	功能说明	
1	FileOutputStream	String name：写入文件名	FileOutputStream	根据参数 name 文件名创建文件输出流对象	

续表

序号	方法名	参数	返回类型	功能说明
2	FileOutputStream	String name：写入文件名；boolean append：是否追加写入，true 表示追加，false 表示覆盖写入	FileOutputStream	根据参数 name 文件名及 append 数据写入方式创建文件输出流对象
3	FileOutputStream	File file：写入文件	FileOutputStream	根据参数 file 文件创建文件输出流对象。
4	FileOutputStream	File file：写入文件；boolean append：是否追加写入，true 表示追加，false 表示覆盖写入	FileOutputStream	根据参数 file 文件及 append 数据写入方式创建文件输出流对

（2）FilterOutputStream 过滤输出流。

FilterOutputStream 过滤输出流提供了用来封装其他的输出流，并为它们提供额外的功能；该类常用的子类有 BufferedOutputStream 缓冲输出流、PipedOutputStream 管道输出字节流、DataOutputStream 数据输出字节流和 PrintStream 打印输出流 4 个子类。

BufferedOutputStream 缓冲输出流

BufferedOutputStream 缓冲输出流为其他输出流提供了缓冲数据的写入操作，该类的描述见表 11-13。

表 11-13 BufferedOutputStream 类描述

类名称	BufferedOutputStream	中文名称	缓冲输出流	类型	实体类
类描述	提供了缓冲数据的写入操作				
序号	方法名	参数	返回类型	功能说明	
1	BufferedOutputStream	OutputStream out：输出流对象	BufferedOutputStream	根据参数 out 创建缓冲输出流对象	
2	BufferedOutputStream	OutputStream out：输出流对象；int size：缓冲字节大小	BufferedOutputStream	根据参数 out 和缓冲大小参数 size 创建缓冲输出流对象	

PipedOutputStream 管道输出流

PipedOutputStream 管道输出流提供了多线程之间进行通信字节数据的读取操作，该类的描述见表 11-14。

表 11-14 PipedOutputStream 类描述

类名称	PipedOutputStream	中文名称	管道输出流	类型	实体类
类描述	提供了多线程之间进行通信字节数据的读取操作				
序号	方法名	参数	返回类型	功能说明	
1	PipedOutputStream	PipedInputStream snk：管道输入流对象	PipedOutputStream	构造方法,根据参数 snk 管道输入流对象构建 PipedOutputStream 对象	
2	PipedOutputStream	无	PipedOutputStream	构造方法，根据默认缓冲大小构建 PipedOutputStream 对象	
5	connect	PipedInputStream snk：管道输入流对象	void	连接到参数 snk 输入流管道	

DataOutputStream 数据输出流

DataOutputStream 数据输出流用来专门写入各种基本数据类型的数据，需要与 DataInputStream 数据输入流配合使用。DataOutputStream 类的描述见表 11-15。

表 11-15　DataOutputStream 类描述

类 名 称	DataOutputStream	中文名称	数据输出流	类　型	实体类
类 描 述	将数据源中的各种基本数据类型的数据进行写入				
序　号	方法名	参　　数	返回类型	功能说明	
1	DataOutputStream	OutputStream src：字节输出流对象	数据输出流对象	构造方法，根据参数 src 字节输出流对象构建 DataOutputStream 对象	
2	writeBoolean	Boolean b1：写入布尔型参数	void	将布尔型参数 b1 写入输出流中	
3	writeByte	byte b：有符号字节参数	void	将字节型参数 b 写入输出流中	
4	writeUnsignedByte	byte b：无符号字节参数	void	将无符号字节参数 b 写入输出流中	
5	writeShort	short s：有符号短整型参数	void	将短整型参数 s 写入输出流中	
6	writeUn-signedShort	short s：无符号短整型参数	void	将无符号短整型参数 s 写入输出流中	
7	writeChar	char c：字符参数	void	将字符型参数 c 写入输出流中	
8	writeInt	int v：整型参数	void	将整型参数 v 写入输出流中	
9	writeLong	Long v：长整型参数	void	将长整型参数 v 写入输出流中	
10	writeFloat	float f：将单精度浮点型参数	void	将单精度浮点型参数 f 写入输出流中	
11	writeDouble	Double d：双精度浮点型参数	void	将双精度浮点型参数 d 写入输出流中	

（3）ByteArrayOutputStream 字节数组输出流。

ByteArrayOutputStream 字节数组输出流提供将字节写入字节数组中，大小不受限制，缓冲会自动增长；该类的描述见表 11-16。

表 11-16　ByteArrayOutputStream 类描述

类 名 称	ByteArrayOutputStream	中文名称	字节数组输出流	类　型	实体类
类 描 述	将字节数组作为流操作，提供了字节数据写入的功能				
序　号	方法名	参　　数	返回类型	功能说明	
1	ByteArrayOutputStream	无	字节数组数据输出流对象	构造方法，构建 ByteArrayOutputStream 对象。	
2	ByteArrayOutputStream	int size：缓存大小	字节数组数据输出流对象	构造方法，构建缓冲大小 size 字节的 ByteArrayOutputStream 对象	
3	writeTo	OutputStream out：输出流对象	void	线程同步操作，将数据写入指定的输出流对象中	

续表

序号	方法名	参数	返回类型	功能说明
4	toByteArray	无	byte[]	返回当前字节数组
5	size	无	int	返回当前字节数组大小

（4）ObjectOutputStream 对象输出流。

ObjectOutputStream 对象输出流是将可序列化对象写入文件中进行存储。读取时需要通过 ObjectInputStream 对象。该类的描述见表 11-17。

表 11-17　ObjectOutputStream 类描述

类名称	ObjectOutputStream	中文名称	对象输出流	类型	实体类
类描述	将对象属性及值写入输出流对象中				
序号	方法名	参数	返回类型	功能说明	
1	ObjectOutputStream	OutputStream out：输出流对象	对象输出流对象	构造方法，根据参数 out 对象构建 ObjectOutputStream 对象	
2	ObjectOutputStream	无	对象输出流对象	构造方法，构建 ObjectOutputStream 对象	
3	writeObject	Object obj	void	将对象 obj 写入流对象中	
4	putFields	无	PutField	返回用于缓冲要写入输出流中的字段的 ObjectOutputStream.PutField 对象	
5	writeFields	无	void	将缓存中的持久层属性写入输出流中	
6	writeBoolean	boolean val	void	将布尔型参数 val 值写入输出流中	
7	writeByte	byte b	void	将有符号参数 b 写入输出流中	
8	writeUnsignedByte	byte b	void	将无符号参数 b 写入输出流中	
9	writeShort	short s	void	将短整型参数 s 写入输出流中	
10	writeUnsignedShort	short s	void	将无符号短整型参数 s 写入输出流中	
11	writeChar	char c	void	将参数 c 写入输出流中	
12	writeInt	int v	void	将整数 v 写入输出流中	
13	writeLong	long v	void	将长整数 v 写入输出流中	

11.2　文件上传

利用本章节的文件上传技术，实现多文件安全上传，并将上传记录写入数据库中；数据库的数据操作采用第 10 章的 JDBC 数据库技术实现。

案例描述

利用 HTML 技术、JavaScript 技术、JSP 技术、JDBC 技术、DBCP 连接池技术及 commons-fileupload 文件上传技术实现多文件上传操作，具体要求如下。

（1）客户端页面上传页面设计如图 11-3 所示。

图 11-3　文件上传页面设计

- 单击图 11-3 中的"上传"按钮，系统提供选择上传文件的对话框，并支持多文件上传；
- 上传文件时，需要在页面显示上传文件的进度及大小，如图 11-4 所示；

图 11-4　文件上传进度

- 已经上传的文件显示在页面上，显示的方式如图 11-3 所示。

（2）上传文件采用 commons-fileupload 组件，显示上传文件的进度；

（3）上传成功的文件信息需要保存在数据库中，每个上传文件的辅助信息至少包括：编号（id）、文件名称（name）、上传时间（createDate）、类型（type）、文件存放地址（address）、大小（size）等。

案例分析

根据案例描述，将文件上传分为页面操作模块、文件上传模块和文件上传数据库操作模块等三个部分，具体每部分设计如图 11-5 所示。

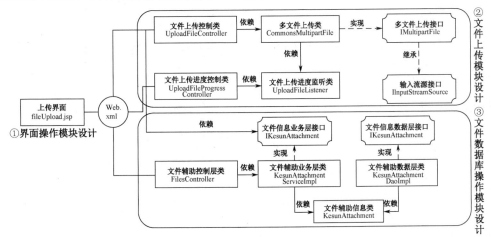

图 11-5　文件上传设计思路

图 11-5 中，详细列出了文件上传功能中各模块之间的关系及每个模块内部功能之间关系。

1. 页面操作模块

页面操作模块由 fileUpload.jsp 页面构成，设计思路：
- 页面中的表单用来实现文件上传，表单中放置文件上传标签，表单隐藏不显示；
- 放置一个普通按钮，按钮上设置文字"上传"，单击该按钮，调用表单中的文件输入标签 change 事件；
- 文件上传时显示的进度采用 DIV 标签显示；
- 文件上传后的显示采用 table 表格标签呈现；文件搜索框使用 input 标签；
- 页面的各种标签样式采用 bootstrap 风格，也可以直接采用原始风格；
- 页面上的数据交互利用 jQuery 的 Ajax 技术实现。

2. 文件上传模块

文件上传模块由输入流源接口、多文件上传接口、多文件上传类、文件上传进度监听类、文件上传控制类和文件上传进度控制类组成，它们之间的关系如图 11-6 所示。

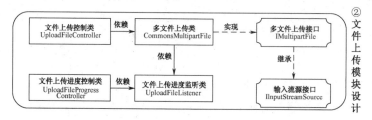

图 11-6 文件上传模块设计

多文件上传接口（IMultipartFile）继承了输入流源接口（IInputStreamSource），多文件上传类（CommonsMultipartFile）实现了多文件上传接口，且依赖文件上传进度监听类（UploadFileListener）。

文件上传控制类（UploadFileController）实现获取页面客户端上传的文件数据，并将文件保存在服务器指定位置，该类依赖多文件上传类（CommonsMultipartFile）。

文件上传进度监听类（UploadFileListener）实现监听文件上传的进度。

文件上传进度控制类（UploadFileProgressController）实现向页面提供上传文件进度数据。具体各种类及接口定义如下：

（1）输入流源接口（IInputStreamSource）。

输入流源接口提供了上传文件输入流数据源，具体的接口描述见表 11-18。

表 11-18 IInputStreamSource 接口描述

类 名 称	IInputStreamSource	中文名称	输入流源接口	类 型	接口
类 描 述	提供了上传文件输入流数据源				
序 号	方 法 名	参 数	返回类型	功能说明	
1	getInputStream	无	InputStream	获取输入流对象	

（2）多文件上传接口（IMultipartFile）。

多文件上传接口实现了文件上传的核心功能，具体接口描述见表 11-19。

表 11-19 IMultipartFile 接口描述

类 名 称	IMultipartFile	中文名称	多文件上传接口	类 型	接口，继承 IInputStreamSource
类 描 述	实现了文件上传的核心功能				
序 号	方法名	参 数	返回类型	功能说明	
1	getName	无	string	用于获得文件上传字段中的文件名	
2	getOriginalFilename	无	string	获取源文件名	
3	getContentType	无	string	获取上传文件的类型	
4	isEmpty	无	boolean	判断上传文件是否为空，如果为空，返回 true，否则返回 false	
5	getSize	无	long	获取上传文件的大小	
6	getBytes	无	Byte[]	获取上传文件的字节	
7	transferTo	File var1：目标文件	void	将当前上传文件内容写入 var1 中	

（3）多文件上传类（CommonsMultipartFile）。

多文件上传类（CommonsMultipartFile）实现了 IMultipartFile 接口，将文件上传的核心功能进行实现，CommonsMultipartFile 类的描述见表 11-20。

表 11-20 CommonsMultipartFile 类的描述

类 名 称	CommonsMultipartFile	中文名称	多文件上传类	类 型	实体类	父类	IMultipartFile 接口
类 描 述	实现多文件上传的功能						
变 量							
序 号	变量名	修饰词	类 型	作 用			
1	fileItem	private final	FileItem	保存当前上传文件			
2	size	private final	long	保存当前上传文件的字节数			
3	preserveFilename	private	boolean	是否保存文件名			
方 法							
序 号	方法名	参 数	返回类型	功能说明			
1	CommonsMultipartFile	FileItem fileItem：上传文件	构造方法	构造方法，根据参数创建多文件上传对象			
2	getFileItem	无	FileItem	获取当前上传文件			
3	setPreserveFilename	boolean preserveFilename：保存文件名	void	设置是否保存文件名			
4	isAvailable	无	boolean	判断是否可以从磁盘或内存中获取上传文件			

（4）文件上传进度监听类（UploadFileListener）。

文件上传进度监听类（UploadFileListener）是实现对上传文件的进度进行监听，通过该类，可以随时获取到当前上传文件的进度、耗时及已上传文件的大小等；该类实现了 ProgressListener 接口，UploadFileListener 类的描述见表 11-21。

表 11-21　UploadFileListener 类的描述

类 名 称	UploadFileListener	中文名称	文件上传进度监听类	类　型	实体类	父类	ProgressListener 接口
类 描 述	实现文件上传进度监控						

变　量			
序　号	变量名	修饰词	类　型	作　用
1	uprate	private	double	上传的速度，每秒字节数
2	percent	private	double	上传完成百分比
3	useTime	private	long	上传已使用的时间
4	upSize	private	long	上传完成的字节数
5	allSize	private	long	总共上传文件的字节数
6	item	private	int	上传第几个文件
7	beginT	private	long	开始上传时间
8	curT	private	long	系统当前时间

方　法			
序　号	方法名	参　数	返回类型	功能说明
1	update	long pBytesRead：已读取字节数；long pContentLength：上传文件总长度；int pItems：上传文件序号	void	获取上传文件最新的上传参数，包括已经读取的字节数、文件总长度及文件序号

（5）文件上传控制类（UploadFileController）。

文件上传控制类是用来接收客户端页面上传的文件，在 CommonsMultipartFile 多文件上传对象及文件附件处理对象协助下最终实现文件上传。文件上传控制类是一个 Servlet 类，它继承了 HTTPServlet 类，该类的描述见表 11-22。

表 11-22　UploadFileController 类的描述

类 名 称	UploadFileController	中文名称	文件上传控制类	类　型	实体类	父类	HTTPServlet 类
类 描 述	在控制层上实现文件上传功能						

方　法			
序　号	方法名	参　数	返回类型	功能说明
1	doUploadFile	CommonsMultipartFile files[]：待上传的文件；HttpServletRequest request：客户端请求对象；HttpServletResponse response：服务器端响应对象	List<String>	将参数 files 上传到服务器上，返回已经上传文件的名称
2	saveAttachment	CommonsMultipartFile file：上传的文件；File newFile：保存到服务器上的文件	void	将上传的文件相关信息保存到数据库中
3	doPost	HttpServletRequest request：客户端请求对象；HttpServletResponse response：服务器端响应对象	void	重写父类中的方法，执行客户端上传文件请求

（6）文件上传进度控制类（UploadFileProgressController）。

文件上传进度控制类是提供给客户端页面用来获取当前文件的上传进度，可以获取到文件的上传速度、大小及耗时等参数，它是一个 Servlet 类，该类的描述见表 11-23。

表 11-23 UploadFileProgressController 类的描述

类名称	UploadFileProgressController	中文名称	文件上传进度控制类	类 型	实体类	父类	HTTPServlet 类
类描述	在控制层上获取文件上传进度参数						
方 法							
序 号	方法名	参 数		返回类型		功能说明	
1	getUploadProgress	HttpServletRequest request：客户端请求对象；		KesunReturn		私有方法，获取文件上传进度参数	
2	doPost	HttpServletRequest request：客户端请求对象；HttpServletResponse response：服务器端响应对象		void		重写父类中的方法，提供客户端上传文件进度数据参数	

3. 文件数据库操作模块

文件数据库操作模块将上传文件的相关信息保存在数据库中，这个模块不会影响文件上传，它可以将上传成功的文件信息保存在数据库中，通过数据库的查找功能，可以查询到完整的文件信息，包括文件原始名称、大小、类型、存放地址等，该模块中的各子模块之间的关系如图 11-7 所示。

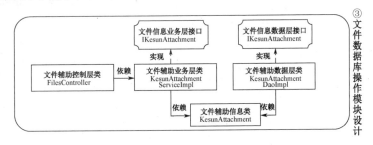

图 11-7 文件数据库操作模块设计

在图 11-7 中，文件数据库操作模块由文件信息数据层接口、文件辅助数据层类、文件信息业务层接口、文件辅助业务层类、文件辅助控制层类和文件辅助信息类 6 个子模块组成。其中，文件信息数据层接口（IKesunAttachment）和文件辅助数据层类（KesunAttachmentDaoImpl）组成了该模块的数据访问层，并且文件辅助数据层类实现了文件信息数据层接口。文件信息业务层接口（IKesunAttachment）和文件辅助业务层类（KesunAttachmentServiceImpl）组成了该模块的业务层，且文件辅助业务层类实现了该层文件信息业务层接口。文件辅助控制层类（FilesController）属于表示层内容。文件辅助信息类（KesunAttachment）分别向该模块的数据访问层、业务层及表示层提供文件辅助信息数据表示支持。

文件数据库操作模块中的数据访问层及业务层设计遵循第 10 章的数据库操作设计，这里不再重复阐述。

（1）文件辅助信息类（KesunAttachment）。

文件辅助信息类（KesunAttachment）是一个 JavaBean 类，用来记录文件的相关基本信息，该类从 AbsObject 继承而来。AbsObject 类前面章节已经介绍过，这里不再重复阐述。KesunAttachment 类的描述见表 11-24。

表 11-24　KesunAttachment 类的描述

类名称	KesunAttachment	中文名称	文件辅助信息类	类　型	实体类	父类	AbsObject
类描述	文件辅助信息基本描述，属于 JavaBean 类						
变　量							
序　号	变量名		修饰词		类　型		作　用
1	type		private		string		上传文件类型
2	size		private		long		上传文件大小，字节单位
3	address		private		string		上传文件存放地址，含完整文件名

（2）文件辅助控制层类（FilesController）。

文件辅助控制层类（FilesController）在控制层上实现上传文件查找功能。该类的描述见表 11-25。

表 11-25　FilesController 类的描述

类名称	FilesController	中文名称	文件辅助控制层类	类　型	实体类	父类	HTTPServlet 类
类描述	在控制层上实现上传文件查找功能。						
方　法							
序　号	方法名		参　数		返回类型		功能说明
1	doPost		HttpServletRequest request：客户端请求对象；HttpServletResponse response：服务器端响应对象		void		重写父类中的方法，在控制层上实现上传文件查找功能

案例实现

编写代码前，需要依赖文件读写操作 commons-io 和文件上传 commons-fileupload 的两个 jar 包，可以从网络下载或从本教材资源包中获取。除此之外，本案例页面使用了 Ajax 数据交互技术，在服务器端 JavaBean 对象与 JSON 之间的转换使用了 fastjson 包，需要从网络上下载。

1. 页面操作模块

（1）页面构成实现代码。

为了提升页面的美观及用户体验感，本案例中的页面部分采用了 BootStrap 框架，脚本使用了 JQuery，这些资源配置放在 init_bootstrap.jsp 页面中，当前页面通过 include 引入即可，上传文件页面名称为 fileUpload.jsp，具体页面实现如下：

```
<%--
    Created by IntelliJ IDEA.
    User: wph-pc
    Date: 2018/12/4
```

```jsp
    Time: 22:03
    To change this template use File | Settings | File Templates.
--%>
<%@ page contentType="text/html;charset=UTF-8" language="java" isELIgnored="false" %>
<%@ include file="../header/init_bootstrap.jsp"%>
<html>
<head>
    <title>文件上传</title>

</head>
<body class="container">
<h2>第11章：多文件上传案例</h2>
<hr>

<div class="panel panel-default">
    <div class="panel-heading panel-primary clearfix">
        <h3>文件上传</h3>
    </div>
    <div class="panel-body">
        <form id="ff" style="display: none;" action="/jspweb/uploadfile" method="post" enctype="multipart/form-data">
            <input type="file" multiple="multiple" name="files" id="txtFile">
        </form>
        <button class="btn btn-primary" id="btnUploadFile" title="单击按钮，实现多文件上传">上传</button>
        <div id="dResult"></div>
        <hr>
        <div class="input-group">
            <input type="text" class="form-control" id="txtFileName" placeholder="请输入文件名称，支持模糊查询">
            <span class="input-group-btn">
                <button class="btn btn-default" type="button" id="btnFindFiles">查找</button>
            </span>
        </div>

        <table class="table">
            <thead>
            <tr>
                <th>序号</th>
                <th>文件名称</th>
                <th>文件类型</th>
                <th>上传日期</th>
                <th>大小</th>
            </tr>
            </thead>
            <tbody id="tFiles">
            </tbody>
        </table>
```

```
        </div>
    </div>
</body>
</html>
```

（2）页面脚本功能实现。

① 页面上传文件查找功能 findFiles。

```
function findFiles() {
    var con=new Object();
    con.fileName=$("#txtFileName").val();
    doData("/files",con,function (data) {
        if (data.obj!=undefined && data.obj!=null){
            $("#tFiles").empty();
            for(var i=0;i<data.obj.length;i++){
                $("#tFiles").append("<tr><td>"+(i+1)+
                    "</td><td>"+data.obj[i].name+"</td><td>"+
                    data.obj[i].type+"</td><td>"+
                    formatDateTimebox(new Date(data.obj[i].createDate))
+"</td><td>"+(data.obj[i].size/1024).toFixed(2)+
                    "<span style='color:deepskyblue'><strong>KB</strong></span></td></tr>")
            }
        }
    },false);
}
```

doData 方法集成了 JQuery 中的 Ajax 技术，已经在前面章节阐述过。formateDateTimebox 方法的功能是实现日期格式转换。

```
function formatDateTimebox (date){
    var y = date.getFullYear();
    var m = date.getMonth()+1;
    var d = date.getDate();
    var h=date.getHours();
    var mm=date.getMinutes();
    var s=date.getSeconds();
    return y+'-'+(m<10?('0'+m):m)+'-'+(d<10?('0'+d):d)+" "+h+":"+mm+":"+s;
}
```

② 页面上传脚本其他相关脚本代码。

```
//时钟事件
var timeEvent=null;
$(function () {
    //默认自动调用已经上传的文件数据
    findFiles();
    //查找按钮执行单击事件
    $("#btnFindFiles").click(function () {
        findFiles();
    });
    //清空上传文件进度提示信息
    $("#dResult").text("");
    //单击"上传"按钮事件
    $("#btnUploadFile").click(function () {
```

```javascript
                //执行文件输入框"单击"事件
                $("#txtFile").click();
        });
        //执行上传文件选中后的事件
        $("#txtFile").change(function () {
                //执行上传功能
                doUploadFile("/jspweb/uploadfile","ff",null,function () {
                        //设置时钟事件，每隔100毫秒获取一次上传文件进度
                        timeEvent= setInterval(function () {
                                //从服务器上获取当前文件的上传进度
                                doData("/uploadfileProgress",null,function (back) {
                                        $("#dResult").text("正在上传第【"+back.obj.item+"】个文件,当前文件上传进度："+
                                        (back.obj.percent*100).toFixed(2)+"%;耗时【"+back.obj.useTime/1000+"】秒"+";上传大小："+
                                        (back.obj.upSize/1024).toFixed(0)+"(kb)/"+(back.obj.allSize/1024).toFixed(0)+"(kb)");
                                        if (back.obj.percent>=1){
                                                //上传文件结束后，清空时钟事件
                                                clearInterval(timeEvent);
                                                //更新页面显示文件
                                                findFiles();
                                        }
                                },false);
                        },100);
                        $("#dResult").text("请稍后，系统开始上传...");
                },function (back) {
                        //上传结束后，显示上传后的信息
                        toastr.info(back.message);
                });
        })

});
```

在以上代码中，执行上传代码的方法 doUploadFile 绑定了 jQuery 的 Ajax 技术，具体定义如下：

```javascript
/*********************************************
 * 功能说明：Ajax文件上传处理技术
 * url地址
 * formId:表单的ID
 * param:json参数对象
 * beforeCallback:处理之前需要做的事件
 * afterCallback:上传返回回调函数
 *********************************************/
function doUploadFile(url,formId,param,beforeCallback,afterCallback)
{
        var con=new FormData($('#'+formId)[0]);
        if (param!=undefined && param!=null)
                jQuery.each(param,function (field, val) {
                        con.append(field,val);
                });
        $.ajax({
```

```
            url:url,
            type: 'POST',
            cache: false,
            data:con,
            processData: false,
            contentType: false,
            dataType:"json",
            beforeSend: function(){
                if ($("#mask").length>0)
                {
                    $("#mask").css("height",$(document).height());
                    $("#mask").css("width",$(document).width());
                   $("#mask img").css("padding-top",window.innerHeight*0.45);
                    $("#mask").show();
                }
                if (beforeCallback!=undefined && beforeCallback!=null)
                    beforeCallback();
            },
            complete:function () {
                if ($("#mask").length>0)
                    $("#mask").hide();
            },
            success : function(data) {
                if (afterCallback!=undefined && afterCallback!=null)
                    afterCallback(data);
            }
        });
    }
```

2. 文件上传模块实现

文件上传模块是服务器接收页面上传文件，保存在服务器指定位置，并协助写入数据库中的核心功能，根据设计，具体实现过程如下：

（1）输入流源接口实现（IInputStreamSource）。

```
package chapter11;
import java.io.IOException;
import java.io.InputStream;
/**
 * 输入流接口
 * Created by wph-pc on 2019/1/14.
 */
public interface IInputStreamSource {
    //获取输入流
    InputStream getInputStream() throws IOException;
}
```

（2）多文件上传接口（IMultipartFile）。

```
package chapter11;
import java.io.File;
import java.io.IOException;
import java.io.InputStream;
/**
```

```
 * 多文件上传接口
 * Created by wph-pc on 2019/1/14.
 */
public interface IMultipartFile extends IInputStreamSource {
    //获取名称
    String getName();
    //获取原文件名称
    String getOriginalFilename();
    //获取文件内容类型
    String getContentType();
    //判断当前文件是否为空,如果为空,返回true,否则返回false
    boolean isEmpty();
    //获取当前文件的大小
    long getSize();
    //获取字节数组
    byte[] getBytes() throws IOException;
    //获取输入流对象
    InputStream getInputStream() throws IOException;
    //将当前文件内容写入参数var1中
    void transferTo(File var1) throws IOException, IllegalStateException;
}
```

（3）多文件上传类（CommonsMultipartFile）。

```
package chapter11;
import org.apache.commons.fileupload.FileItem;
import org.apache.commons.fileupload.FileUploadException;
import org.apache.commons.fileupload.disk.DiskFileItem;

import java.io.File;
import java.io.IOException;
import java.io.InputStream;
import java.io.Serializable;

/**
 * 多文件上传类
 * Created by wph-pc on 2019/1/14.
 */
public class CommonsMultipartFile implements IMultipartFile,Serializable {
    private final FileItem fileItem;
    //声明最终长度变量size
    private final long size;
    //用于处理文件名,布尔型变量
    private boolean preserveFilename = false;

    public CommonsMultipartFile(FileItem fileItem) {
        this.fileItem = fileItem;
        this.size = this.fileItem.getSize();
    }

    public final FileItem getFileItem() {
        return this.fileItem;
    }
```

```java
public void setPreserveFilename(boolean preserveFilename) {
    this.preserveFilename = preserveFilename;
}

public String getName() {
    return this.fileItem.getFieldName();
}
/*获取上传文件的源文件名称*/
public String getOriginalFilename() {
    String filename = this.fileItem.getName();
    if(filename == null) {
        return "";
    } else if(this.preserveFilename) {
        return filename;
    } else {
        int unixSep = filename.lastIndexOf("/");
        int winSep = filename.lastIndexOf("\\");
        int pos = winSep > unixSep?winSep:unixSep;
        return pos != -1?filename.substring(pos + 1):filename;
    }
}

public String getContentType() {
    return this.fileItem.getContentType();
}

public boolean isEmpty() {
    return this.size == 0L;
}

public long getSize() {
    return this.size;
}

public byte[] getBytes() {
    if(!this.isAvailable()) {
        throw new IllegalStateException("文件已经移动,不能再次读取!");
    } else {
        byte[] bytes = this.fileItem.get();
        return bytes != null?bytes:new byte[0];
    }
}

public InputStream getInputStream() throws IOException {
    if(!this.isAvailable()) {
        throw new IllegalStateException("文件已经移动,不能再次读取!");
    } else {
        InputStream inputStream = this.fileItem.getInputStream();
        return inputStream != null?inputStream: null;
    }
```

```java
        }
        /*将当前文件fileItem的数据写入dest文件中*/
        public void transferTo(File dest) throws IOException, IllegalStateException {
            if(!this.isAvailable()) {
                throw new IllegalStateException("文件已经传送完成,不能再次传送.");
            } else if(dest.exists() && !dest.delete()) {
                throw new IOException("目标文件[" + dest.getAbsolutePath() + "]已经存在且不能删除.");
            } else {
                try {
                    //源文件数据写入目标文件
                    this.fileItem.write(dest);
                } catch (FileUploadException var3) {
                    throw new IllegalStateException(var3.getMessage(), var3);
                } catch (IllegalStateException var4) {
                    throw var4;
                } catch (IOException var5) {
                    throw var5;
                } catch (Exception var6) {
                    throw new IOException("文件写入失败", var6);
                }
            }
        }
        /*判断当前文件是否可以从内存或磁盘上获取,若可以获取,返回true,否则返回false*/
        protected boolean isAvailable() {
            return this.fileItem.isInMemory()?true:(this.fileItem instanceof DiskFileItem?((DiskFileItem)this.fileItem).getStoreLocation().exists():this.fileItem.getSize() == this.size);
        }
        /*获取当前上传文件存放的位置描述*/
        public String getStorageDescription() {
            return this.fileItem.isInMemory()?"内存中":(this.fileItem instanceof DiskFileItem?"at [" + ((DiskFileItem)this.fileItem).getStoreLocation().getAbsolutePath() + "]":"磁盘上");
        }
    }
```

(4) 文件上传进度监听类（UploadFileListener）。

```java
package chapter11;
import org.apache.commons.fileupload.ProgressListener;
/**
 * 上传文件进度监听
 * Created by wph-pc on 2019/1/14.
 */
public class UploadFileListener implements ProgressListener{
    //上传的速度,每秒字节数
    private double upRate = 0.0;
    //上传完成百分比
    private double percent = 0.0;
    //上传已用去的时间
```

```java
        private long useTime = 0;
        //上传完成的字节数
        private long upSize = 0;
        //总共上传文件的字节数
        private long allSize = 0;
        //上传第几个文件
        private int item;
        //开始上传时间
        private long beginT =System.currentTimeMillis();
        //系统当前时间
        private long curT =System.currentTimeMillis();
        /*
        *文件上传时调用的方法
        * @param pBytesRead:已经读取的字节数
        * @param pContentLength:上传文件的长度
        * @param pItems:上传第几个文件
        */
        public void update(long pBytesRead, long pContentLength, int pItems) {
            curT =System.currentTimeMillis();
            item = pItems;
            allSize = pContentLength;   //byte
            upSize = pBytesRead;    //byte
            useTime = curT-beginT;   //ms
            if(useTime != 0)
                upRate =pBytesRead/useTime;   //byte/ms
            else
                upRate =0.0;
            if(pContentLength == 0)
                return;
            percent =(double)pBytesRead/(double)pContentLength;   //百分比
            System.out.println("上传进度百分比: "+percent*100+"%"+";
耗时: "+useTime);

        }
        public long getAllSize() {
            return allSize;
        }
        public void setAllSize(long allSize) {
            this.allSize = allSize;
        }
        public long getBeginT() {
            return beginT;
        }
        public void setBeginT(long beginT) {
            this.beginT = beginT;
        }
        public long getCurT() {
            return curT;
        }
        public void setCurT(long curT) {
            this.curT = curT;
```

```java
    }
    public int getItem() {
        return item;
    }
    public void setItem(int item) {
        this.item = item;
    }
    public double getPercent() {
        return percent;
    }
    public void setPercent(double percent) {
        this.percent = percent;
    }
    public double getUpRate() {
        return upRate;
    }
    public void setUpRate(double upRate) {
        this.upRate = upRate;
    }
    public long getUpSize() {
        return upSize;
    }
    public void setUpSize(long upSize) {
        this.upSize = upSize;
    }
    public long getUseTime() {
        return useTime;
    }
    public void setUseTime(long useTime) {
        this.useTime = useTime;
    }
}
```

（5）文件上传控制类（UploadFileController）。

```java
package chapter11;
import business.entity.KesunReturn;
import chapter11.entity.KesunAttachment;
import chapter11.service.impl.KesunAttachmentServiceImpl;
import com.alibaba.fastjson.JSON;
import com.alibaba.fastjson.JSONObject;
import org.apache.commons.fileupload.FileItem;
import org.apache.commons.fileupload.FileItemFactory;
import org.apache.commons.fileupload.FileUploadException;
import org.apache.commons.fileupload.ProgressListener;
import org.apache.commons.fileupload.disk.DiskFileItemFactory;
import org.apache.commons.fileupload.Servlet.ServletFileUpload;
import util.JSONTool;

import javax.Servlet.ServletContext;
import javax.Servlet.ServletException;
import javax.Servlet.http.HttpServlet;
import javax.Servlet.http.HttpServletRequest;
```

```java
import javax.Servlet.http.HttpServletResponse;
import java.io.File;
import java.io.IOException;
import java.text.SimpleDateFormat;
import java.util.ArrayList;
import java.util.Date;
import java.util.List;
import java.util.UUID;

/**
 * 文件上传处理类
 * Created by wph-pc on 2019/1/14.
 */
public class UploadFileController extends HttpServlet {
    /*
    *文件上传操作
    * @param:files[]上传的文件组；
    * @param:request:客户端请求对象；
    * @param:response:服务器端请求对象；
    * @return:返回以上上传成功的文件名
    * */
    private List<String> doUploadFile(CommonsMultipartFile files[], HttpServletRequest request, HttpServletResponse response)
    {
        response.setContentType("text/html;charset=UTF-8");
        if(files==null || files.length==0) return null;
        //获取上传文件保存在服务器上的逻辑地址
        String path=request.getSession().getServletContext().getRealPath("/uploadfiles");
        /*判断目录是否存在，如果不存在，则创建*/
        File dir = new File(path);
        if (!dir.exists())
            dir.mkdirs();
        List<String> lFiles=new ArrayList<String>();//已经上传成功的文件名
        for (int i = 0; i < files.length; i++) {
            // 获得原始文件名
            String fileName = files[i].getOriginalFilename();
            if (fileName.trim().equals("")) continue;
            // 新文件名
            String newFileName = UUID.randomUUID().toString().replace("-","")+fileName.substring(fileName.lastIndexOf("."));
            if (!files[i].isEmpty()) {
                try {
                    //创建新文件
                    File newFile=new File(path+"/"+newFileName);
                    //将上传文件中的数据发送到创建的新文件中
                    files[i].transferTo(newFile);
                    //将上传的文件信息转成附件保存到数据库
                    saveAttachment(files[i],newFile);

                } catch (Exception e) {
```

```java
                e.printStackTrace();
                return null;
            }
        }
        lFiles.add(newFileName);
    }
    return lFiles;
}
/*
*将参数file转换成attachment保存到数据库
* */
private void saveAttachment(CommonsMultipartFile file,File newFile){
    /*将上传的文件信息写入数据库*/
    KesunAttachment attachment=new KesunAttachment();
    //设置自动编号ID
    attachment.setId(UUID.randomUUID().toString().replace("-",""));
    attachment.setSize(file.getSize());
    attachment.setName(file.getOriginalFilename().substring(0,file.getOriginalFilename().indexOf(".")));
    attachment.setType(file.getOriginalFilename().substring(file.getOriginalFilename().indexOf(".")+1));
    attachment.setAddress(newFile.getName());

    //创建附件业务层对象
    KesunAttachmentServiceImpl attachBll=new KesunAttachmentServiceImpl();
    attachBll.setModel(attachment);
    //执行附件新增功能
    attachBll.add();
}
@Override
public void doPost(HttpServletRequest request,HttpServletResponse response)
        throws ServletException, IOException {
    //设置客户端字符编码
    request.setCharacterEncoding("utf-8");
    //设置返回页面上的JSON格式
    response.setContentType("application/json;charset=utf-8");
    //设置返回对象
    KesunReturn back=new KesunReturn();
    if (!ServletFileUpload.isMultipartContent(request)){
        back.setMessage("系统检测到上传文件格式错误！");
        back.setCode("0");
        back.setObj(null);
        response.getWriter().write(JSON.toJSONString(back));
        return;
    }
    //创建文件项工厂对象
    FileItemFactory fileItemFactory = new DiskFileItemFactory();
    //创建文件上传对象
    ServletFileUpload ServletFileUpload = new ServletFileUpload
```

```
(fileItemFactory);
                // 设置上传的单个文件的最大字节数为200MB
                ServletFileUpload.setFileSizeMax(1024*1024*200);
                //设置整个表单的最大字节数为1GB
                ServletFileUpload.setSizeMax(1024*1024*1024);
                //设置文件上传进度监听
                UploadFileListener progress=new UploadFileListener();
                //绑定上传进度对象
                ServletFileUpload.setProgressListener(progress);
                //将progress进度对象写入session
                request.getSession().setAttribute("progress",progress);
                try {
                    //获取上传的文件
                    List<FileItem> list = ServletFileUpload.parseRequest(request);
                    /*判断文件的有效性*/
                    if (list==null || list.size()==0){
                        back.setMessage("系统没有获取到上传文件！");
                        back.setCode("-1");
                        back.setObj(null);
                        response.getWriter().write(JSON.toJSONString(back));
                        return;
                    }
                    //将list中的文件转换成CommonsMultipartFile类型
                    CommonsMultipartFile[] source=new CommonsMultipartFile[list.size()];
                    for(int i=0;i<list.size();i++){
                        source[i]=new CommonsMultipartFile(list.get(i));
                    }
                    //已经完成的上传文件
                    List<String> lFinished=doUploadFile(source,request,response);
                    if(lFinished!=null && lFinished.size()>0){
                        back.setObj(lFinished);
                        back.setMessage("文件上传成功！");
                        back.setCode("1");
                    }else{
                        back.setObj(null);
                        back.setMessage("文件上传失败！");
                        back.setCode("-2");
                    }
                    response.getWriter().write(JSON.toJSONString(back));
                } catch (FileUploadException e) {
                    back.setCode("-3");
                    back.setMessage(e.getMessage());
                    back.setObj(null);
                    response.getWriter().write(JSON.toJSONString(back));
                }
            }
        }
```

文件上传控制类必须在 web.xml 中进行 Servlet 节点配置，具体配置如下：

```
<Servlet>
```

```xml
    <Servlet-name>uploadfile</Servlet-name>
    <Servlet-class>chapter11.UploadFileController</Servlet-class>
</Servlet>
```
文件上传控制类访问地址映射配置如下:
```xml
<Servlet-mapping>
    <Servlet-name>uploadfile</Servlet-name>
    <url-pattern>/uploadfile</url-pattern>
</Servlet-mapping>
```
(6) 文件上传进度控制类（UploadFileProgressController）。
```java
package chapter11;
import business.entity.KesunReturn;
import com.alibaba.fastjson.JSON;
import javax.Servlet.ServletException;
import javax.Servlet.http.HttpServlet;
import javax.Servlet.http.HttpServletRequest;
import javax.Servlet.http.HttpServletResponse;
import java.io.IOException;

/**
 * 获取文件的上传进度
 * Created by wph-pc on 2019/1/16.
 */
public class UploadFileProgressController extends HttpServlet {
    @Override
    public void doPost(HttpServletRequest request, HttpServletResponse response)
            throws ServletException, IOException {
        //设置客户端字符编码
        request.setCharacterEncoding("utf-8");
        //设置返回页面上的JSON格式
        response.setContentType("application/json;charset=utf-8");
        response.getWriter().write(JSON.toJSONString(getUploadProgress(request)));
    }
    private KesunReturn getUploadProgress(HttpServletRequest request){
        KesunReturn back=new KesunReturn();
        Object obj=request.getSession().getAttribute("progress");
        if (obj==null || obj instanceof UploadFileListener==false){
            back.setCode("0");
            back.setMessage("系统没有获取到进度对象");
        }else
        {
            back.setCode("1");
            back.setMessage("系统已获取到进度对象");
        }
        back.setObj(obj);
        return back;
    }
}
```
要使用文件上传进度控制类，需要在 web.xml 中配置 Servlet 节点:
```xml
<Servlet>
```

```xml
    <Servlet-name>uploadfileProgress</Servlet-name>
    <Servlet-class>chapter11.UploadFileProgressController</Servlet-class>
</Servlet>
```

文件上传进度控制类 Servlet 地址映射节点配置：

```xml
<Servlet-mapping>
    <Servlet-name>uploadfileProgress</Servlet-name>
    <url-pattern>/uploadfileProgress</url-pattern>
</Servlet-mapping>
```

3. 文件数据库操作模块实现

文件数据库操作模块是将已经上传到服务器上的文件信息写入数据库中，每个上传的文件，在服务器上保存的名称都是唯一码，通过服务器上的文件名无法详细读取文件本身的辅助信息。因此上传文件时，将原来文件本身的辅助信息保存在数据库中，提取时就更容易辨别文件信息。具体实现根据文件数据库操作模块设计。

（1）文件辅助信息类（KesunAttachment）。

文件辅助信息类 KesunAttachment 用来描述文件基本信息 JavaBean 类，具体定义如下：

```java
package chapter11.entity;
import business.entity.AbsObject;
/**
 * 文件上传附件
 * Created by wph-pc on 2018/8/19.
 */
public class KesunAttachment extends AbsObject {
    private String type="";//附件类型
    private long size=0;//附件大小
    private String address="";//附件存放地址
    public String getType() {
        return type;
    }
    public void setType(String type) {
        this.type = type;
    }
    public long getSize() {
        return size;
    }
    public void setSize(long size) {
        this.size = size;
    }
    public String getAddress() {
        return address;
    }
    public void setAddress(String address) {
        this.address = address;
    }
}
```

（2）文件信息数据层接口（IKesunAttachment）。

文件信息数据层接口（IKesunAttachment）从 IDoData 接口继承，该接口定义了数据库的常规操作标准，在第 10 章已经详细阐述过。文件信息数据层接口定义如下：

```
package chapter11.dao;

import business.dao.IDoData;

/**
 * Created by wph-pc on 2019/1/17.
 */
public interface IKesunAttachment extends IDoData {
}
```

（3）文件辅助数据层类（KesunAttachmentDaoImpl）。

文件辅助数据层类（KesunAttachmentDaoImpl）实现了文件信息数据层接口IKesunAttachment，主要实现了数据保存及查找功能，具体实现代码如下：

```
package chapter11.dao.impl;
import business.entity.AbsObject;
import chapter11.dao.IKesunAttachment;
import chapter11.entity.KesunAttachment;
import java.text.ParseException;
import java.text.SimpleDateFormat;
import java.util.*;
/**
 * 文件辅助数据层类
 * Created by wph-pc on 2019/1/17.
 */
public class KesunAttachmentDaoImpl implements IKesunAttachment {
    /*将参数obj转成KesunAttachment文件辅助信息对象*/
    private KesunAttachment getAttachment(AbsObject obj){
        //判断obj是否是文件附件类型
        if (obj==null || obj instanceof KesunAttachment ==false) return null;
        //将obj转换成目标类型
        KesunAttachment attachment=(KesunAttachment)obj;
        return attachment;
    }
    @Override
    public int add(AbsObject obj) {
        //将对象obj转换成附件类型
        KesunAttachment model=getAttachment(obj);
        //判断model是否符合条件
        if (model==null) return 0;
        model.setCreateDate(new Date());
        SimpleDateFormat sdf = new SimpleDateFormat("yyyy-MM-dd HH:mm:ss");
        //将model对象组成新增指令
        String cmd="insert into attachment (id,name,createDate,description,type,address,size) values ('"+
                model.getId()+"','"+model.getName()+"','"+sdf.format(model.getCreateDate())+"','"+model.getDescription()+"','"+
                model.getType()+"','"+model.getAddress()+"',"+model.getSize()+")";
        //创建数据库访问对象
        chapter10.DBHelper command=chapter10.DBHelper.getInstance();
```

```java
            //执行附件新增操作
            return command.command(cmd);
    }

    @Override
    public int edit(AbsObject obj) {
        //将对象obj转换成附件类型
        KesunAttachment model=getAttachment(obj);
        //判断model是否符合条件
        if (model==null) return 0;
        //将model对象组成修改指令
        String cmd="update attachment set name='"+model.getName()+
                "',createDate='"+model.getCreateDate().toString()+
                "',description='"+model.getDescription()+
                "',type='"+model.getType()+"',address='"+model.getAddress()+
                "',size="+model.getSize()+" where id='"+model.getId()+"'";
        //创建数据库访问对象
        chapter10.DBHelper command=chapter10.DBHelper.getInstance();
        //执行附件修改操作
        return command.command(cmd);
    }

    @Override
    public int del(AbsObject obj) {
        //将对象obj转换成附件类型
        KesunAttachment model=getAttachment(obj);
        //判断model是否符合条件
        if (model==null) return 0;
        //将model对象组成删除指令
        String cmd="delete from attachment where id='"+model.getId()+"'";
        //创建数据库访问对象
        chapter10.DBHelper command=chapter10.DBHelper.getInstance();
        //执行附件删除操作
        return command.command(cmd);
    }

    @Override
    public int changeStatus(AbsObject obj) {
        return 0;
    }

    /*方法名：getCondition,获取查询条件
    *作用：将condition参数转成字符串查询参数，参数condition的key必须在source中视为有效
    *@param condtion:查询条件
    *@param source:有效key
    *@return 如果转换成功，返回有效的查询字符串，否则返回null*/
    private String getCondition(Map<String,Object> condition,List<String> source){
        //判断条件的有效性
```

```java
        if (condition==null || condition.size()==0 || source==null || source.size()==0) return null;
        //定义变量保存的有效条件
        StringBuilder sb=new StringBuilder();
        sb.append(" where ");
        //获取condition中的key集合
        Iterator<String> keySources=condition.keySet().iterator();
        while (keySources.hasNext()){
            //获取当前key名称
            String key=keySources.next();
            //判断当前key是否在source中
            if (source.contains(key))
                if (condition.get(key) instanceof Integer)
                    sb.append( key+"="+condition.get(key).toString()+" and ");
                else
                    sb.append( key+" like '%"+condition.get(key).toString()+"%' and ");
        }
        /*处理变量sb中的数据*/
        if (sb.toString().equals(" where ")) return null;
        //获取最后一个and出现的位置
        int lastIndex=sb.toString().lastIndexOf("and");
        if (lastIndex>0)
            return sb.toString().substring(0,lastIndex-1).trim();
        else
            return sb.toString();
    }
    /*设置有限查询列信息*/
    private List<String> checkColomnSource(){
        List<String> source=new ArrayList<>();
        source.add("id");
        source.add("name");
        source.add("type");
        return source;
    }
    @Override
    public List<Map<String, Object>> findResult(Map<String, Object> condition) {
        //对条件condition进行处理,转成有效的字符串查询条件
        String queryCondition=getCondition(condition,checkColomnSource());
        //组建查询指令
        String sql="select * from attachment ";
        if (queryCondition!=null && !"".equals(queryCondition.trim()))
            sql+=" "+queryCondition;
        //创建数据库访问对象DBHelper
        chapter10.DBHelper dbHelper= chapter10.DBHelper.getInstance();
        //数据库执行附件查询
        return dbHelper.find(sql+" order by createDate desc");
    }
```

```java
@Override
public List<AbsObject> find(Map<String, Object> condition) {
    //查找符合条件的附件信息
    List<Map<String, Object>> result=findResult(condition);
    if (result==null || result.size()==0) return null;
    //将result转换成List<AbsObject>
    List<AbsObject> lObjs=new ArrayList<>();

    Iterator<Map<String,Object>> source=result.iterator();
    while(source.hasNext()){
        Map<String,Object> temp=source.next();
        KesunAttachment attach=new KesunAttachment();
        if (temp.get("id")!=null)
            attach.setId(temp.get("id").toString());
        if (temp.get("name")!=null)
            attach.setName(temp.get("name").toString());
        if (temp.get("createDate")!=null) {
            SimpleDateFormat sdf = new SimpleDateFormat("yyyy-MM-dd HH:mm:ss");
            try {
                attach.setCreateDate(sdf.parse(temp.get("createDate").toString()));
            } catch (ParseException e) {
                attach.setCreateDate(null);
                e.printStackTrace();
            }
        }
        if (temp.get("address")!=null)
            attach.setAddress(temp.get("address").toString());
        if (temp.get("size")!=null)
            attach.setSize(Integer.valueOf(temp.get("size").toString()));
        if (temp.get("type")!=null)
            attach.setType(temp.get("type").toString());
        lObjs.add(attach);
    }
    return lObjs;
}

@Override
public AbsObject getMe(String id) {
    Map<String,Object> cons=new HashMap<>();
    //附件账号ID作为查询条件
    cons.put("id",id);
    //根据ID查询附件
    List<AbsObject> objs=find(cons);
    if (objs!=null && objs.size()>0 && objs.get(0) instanceof KesunAttachment)
        return objs.get(0);
    else
        return null;
```

```
    }
}
```

(4) 文件信息业务层接口（IKesunAttachment）。

文件信息业务层接口（IKesunAttachment）是一个扩展接口，定义如下：

```
package chapter11.service;
/**
 * 文件信息业务层接口
 * Created by wph-pc on 2019/1/17.
 */
public interface IKesunAttachment {
}
```

(5) 文件辅助业务层类（KesunAttachmentServiceImpl）。

文件辅助业务层类（KesunAttachmentServiceImpl）是上传的文件基本信息实现提交给数据访问层进行数据库的相关操作，它实现了文件信息业务层接口 IKesunAttachment，并且继承了 KesunSuperService 业务层超级类，该类在第 10 章已经详细阐述过。KesunAttachmentServiceImplement 具体定义如下：

```
package chapter11.service.impl;
import business.dao.IDoData;
import business.service.KesunSuperService;
import chapter11.entity.KesunAttachment;
import chapter11.service.IKesunAttachment;
/**
 * 文件辅助业务层类
 * Created by wph-pc on 2019/1/17.
 */
public class KesunAttachmentServiceImpl extends KesunSuperService implements IKesunAttachment{
    public KesunAttachmentServiceImpl(){
        setModel(new KesunAttachment());
    }
    @Override
    public IDoData getDAO() {
        return new chapter11.dao.impl.KesunAttachmentDaoImpl();
    }
}
```

(6) 文件辅助控制层类（FilesController）。

文件辅助控制层类（FilesController）是一个 Servlet 类，提供了上传文件的查询功能，具体定义如下：

```
package chapter11;
import chapter11.service.impl.KesunAttachmentServiceImpl;
import com.alibaba.fastjson.JSONObject;
import util.JSONTool;
import javax.Servlet.RequestDispatcher;
import javax.Servlet.ServletException;
import javax.Servlet.http.HttpServlet;
import javax.Servlet.http.HttpServletRequest;
import javax.Servlet.http.HttpServletResponse;
import java.io.IOException;
```

```java
    import java.util.HashMap;
    import java.util.Map;
    /**
     * 文件辅助控制层类
     * Created by wph-pc on 2019/1/18.
     */
    public class FilesController extends HttpServlet {
        private static final long serialVersionUID = 6765085208899952417L;
        public void doGet(HttpServletRequest request, HttpServletResponse response)
                throws ServletException, IOException {
           doPost(request, response);
        }

        public void doPost(HttpServletRequest request, HttpServletResponse response)
                throws ServletException, IOException {
           /*设置编码格式及响应类型*/
           request.setCharacterEncoding("utf-8");
           response.setContentType("application/json");
           response.setCharacterEncoding("utf-8");

           //将获取的客户端请求的参数转成JSONObject类型
           JSONObject jsonObject= JSONTool.GetRequestJSON(request);
           //获取文件名参数
           String fileName=jsonObject.getString("fileName");
           /*将文件查询参数转成Map结构*/
           Map<String,Object> cons=new HashMap<String,Object>();
           if (fileName!=null)
              cons.put("name",fileName);
           //创建文件附件业务对象
           KesunAttachmentServiceImpl bll=new KesunAttachmentServiceImpl();
           //查询附件并输出到页面
           response.getWriter().write(JSONObject.toJSONString(bll.findForMap(cons)));
           //关闭输出流对象
           response.getWriter().close();
        }
    }
```

文件辅助控制类需要在 web.xml 中进行注册，配置 Servlet 节点如下：

```xml
<Servlet>
    <Servlet-name>files</Servlet-name>
    <Servlet-class>chapter11.FilesController</Servlet-class>
</Servlet>
```

文件辅助控制层类在 web.xml 中的 Servlet 地址映射配置：

```xml
<Servlet-mapping>
    <Servlet-name>files</Servlet-name>
    <url-pattern>/files</url-pattern>
</Servlet-mapping>
```

运行结果

运行 fileUpload.jsp 页面，其运行结果如图 11-8 所示。

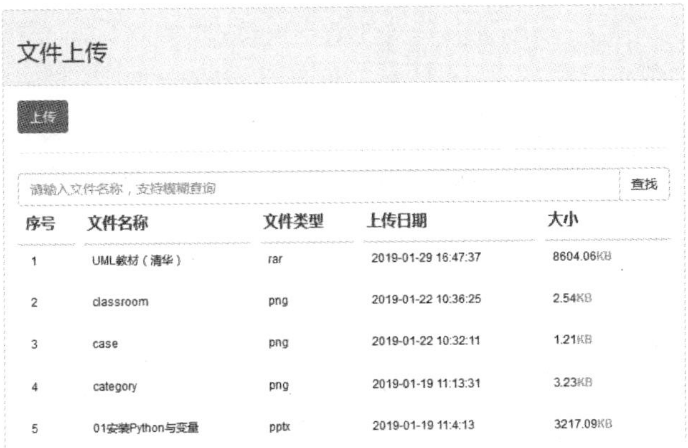
文件上传

上传

请输入文件名称，支持模糊查询 查找

序号	文件名称	文件类型	上传日期	大小
1	UML教材（清华）	rar	2019-01-29 16:47:37	8604.06KB
2	classroom	png	2019-01-22 10:36:25	2.54KB
3	case	png	2019-01-22 10:32:11	1.21KB
4	category	png	2019-01-19 11:13:31	3.23KB
5	01安装Python与变量	pptx	2019-01-19 11:4:13	3217.09KB

图 11-8　文件上传运行结果

单击图 11-8 中的“上传”按钮，系统弹出选择文件的对话框，用户选择需要上传的文件，在页面会显示上传文件的进度。上传完毕后，上传的文件会显示在表格中，也可以输入文件名称，支持模糊查询，单击“查找”按钮即可实现文件查询，相关的文件信息全部来自于数据库，而实际存放在服务器上的文件名称是一个随机码。

11.3　文件下载

文件下载是将服务器上的文件下载到客户机上。本节案例是将服务器上的文件进行下载，但是不能直接通过链接下载，需要通过 IO 流的方式。

案例描述

利用上传文件的结果，将已经上传的文件下载到客户端，文件下载页面设计如图 11-9 所示。

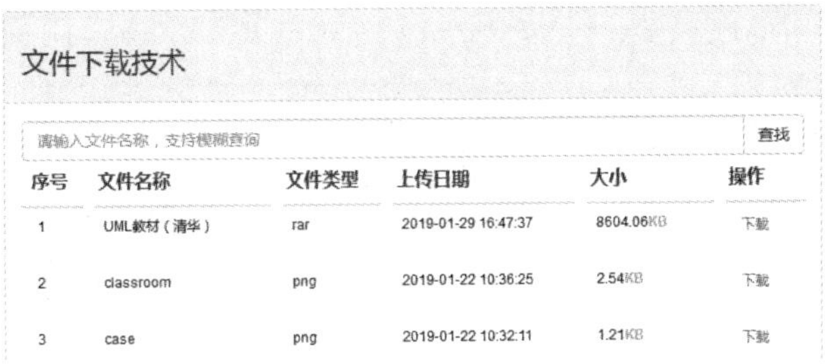
文件下载技术

请输入文件名称，支持模糊查询 查找

序号	文件名称	文件类型	上传日期	大小	操作
1	UML教材（清华）	rar	2019-01-29 16:47:37	8604.06KB	下载
2	classroom	png	2019-01-22 10:36:25	2.54KB	下载
3	case	png	2019-01-22 10:32:11	1.21KB	下载

图 11-9　文件下载页面设计

图 11-9 中，文件下载页面与案例文件上传页面基本一致，不同之处在于每个下载文件最后一列增加了“下载”按钮，单击该按钮执行下载。

具体要求：
- 文件查找及显示功能与文件上传一致；
- 在每个文件后面增加一个"下载"按钮，下载按钮绑定的只是文件存储在数据库中的 ID，与保存在服务器中的文件名不一致；
- 通过表格中每行文件的 ID 查找到文件属性 address 的值，该值即是存放在服务器上的文件名，根据该值进行 IO 流数据写入，写入完成后实现下载；
- 不能通过"下载"按钮直接绑定服务器上的下载文件链接地址进行下载。

案例分析

直接通过"<a>"链接下载很不安全，会将下载地址暴露在网络上。因此，隐藏要下载的文件地址相对较为安全。本章通过 IO 流的方式进行下载。

文件下载流程设计如图 11-10 所示。

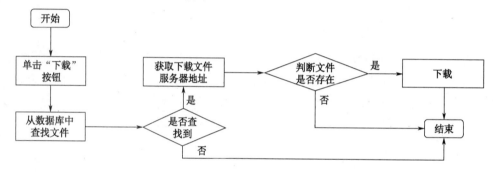

图 11-10 文件下载流程设计

从图 11-10 中可以看出，单击图 11-9 中的"下载"按钮，系统先从数据库中查找当前下载的文件，如果没有查找到，操作结束；如果查找到了，返回当前文件对象，从文件对象中获取下载文件服务器地址，并判断当前地址的文件是否存在，如果存在则进行下载，否则结束操作。

下载文件查找在文件上传案例中已经阐述，此处不再重复讲述。下载功能由 DownloadFileController 类完成，该类是一个 Servlet 类，其描述见表 11-26。

表 11-26 DownloadFileController 类的描述

类 名 称	DownloadFileController	中文名称	文件下载类	类 型	实体类	父类	HttpServlet
类 描 述	实现文件下载功能						
方　法							
序　号	方 法 名	参　数		返回类型	功能说明		
1	getAttachment	String id：下载文件编号		KesunAttachment	私有方法，根据参数 id 从数据库中获取文件信息		
2	doPost	HttpServletRequest request：请求内置对象；HttpServletResponse response：服务器端响应对象		void	采用 IO 流形式实现文件下载。		

案例实现

文件下载主要是通过文件下载类 DownLoadFileController 实现的，该类是一个 Servlet 类，必须在 web.xml 中进行配置。

1. DownLoadFileController 类

```java
package chapter11;

import business.entity.KesunReturn;
import chapter11.entity.KesunAttachment;
import chapter11.service.impl.KesunAttachmentServiceImpl;
import com.alibaba.fastjson.JSONObject;
import util.JSONTool;

import javax.Servlet.ServletException;
import javax.Servlet.http.HttpServlet;
import javax.Servlet.http.HttpServletRequest;
import javax.Servlet.http.HttpServletResponse;
import java.io.*;
import java.net.URLEncoder;

/**
 * 采用文件流方式进行文件下载
 * Created by wph-pc on 2019/1/16.
 */
public class DownloadFileController extends HttpServlet {
    public void doGet(HttpServletRequest request, HttpServletResponse response)
            throws ServletException, IOException {
        doPost(request, response);
    }
    /*根据附件ID获取附件*/
    private KesunAttachment getAttachment(String id){
        //创建附件业务层对象
        KesunAttachmentServiceImpl bll=new KesunAttachmentServiceImpl();
        //设置获取附件对象ID
        bll.getModel().setId(id);
        //获取文件附件
        KesunReturn back=bll.getMe();
        //如果没有获取到，返回null
        if (back.getObj()==null || back.getObj() instanceof KesunAttachment==false) return null;
        return (KesunAttachment)back.getObj();
    }

    public void doPost(HttpServletRequest request, HttpServletResponse response)
            throws ServletException, IOException {
        String basePath = request.getSession().getServletContext().getRealPath ("/uploadfiles");
        //创建返回对象
```

```java
KesunReturn back=new KesunReturn();
/*获取下载文件的ID*/
String fileId=request.getParameter("fileID");
if (fileId==null){
    back.setMessage("系统没有获取到您下载的文件ID");
    back.setCode("0");
    back.setObj(null);
    response.getWriter().write(JSONObject.toJSONString(back));
    return;
}
/*根据文件的ID从数据库中获取文件名*/
KesunAttachment attachment=getAttachment(fileId);
//根据附件对象组成下载文件的地址
String filePath = basePath + File.separator + attachment.getAddress();
/*判断文件是否存在*/
File file=new File(filePath);
if (file.exists()==false){
    back.setCode("-1");
    back.setMessage("您要下载的文件已经不存在！");
    back.setObj(null);
    response.getWriter().write(JSONObject.toJSONString(back));
    return;
}
//设置头响应,控制浏览器下载该文件,并解决文件名中含有中文不显示或乱码问题
response.addHeader("Content-Disposition", "attachment;filename*=UTF-8''" + URLEncoder.encode((attachment.getName()+"."+attachment.getType()), "UTF-8").replaceAll("\\+", "%20"));
//定义输入流变量,用来读取下载源文件数据
InputStream is = null;
//定义输出流变量,用来向浏览器输出下载文件
OutputStream os = null;
//定义缓冲输入流变量
BufferedInputStream bis = null;
//定义缓冲输出流变量
BufferedOutputStream bos = null;
//根据下载文件地址创建输入流对象,并赋值给is输入流变量
is = new FileInputStream(new File(filePath));
//根据输入流变量is创建缓冲输入流对象,并赋值给bis变量
bis = new BufferedInputStream(is);
//通过response内置对象获取输出流对象,并赋值给os输出流变量
os = response.getOutputStream();
//创建缓冲输出流对象,并赋值给输出流变量bos
bos = new BufferedOutputStream(os);
//定义字节数组,作为读取数据缓冲用
byte[] b = new byte[1024];
//定义变量,保存每次读取有效字节的个数
int len = 0;
//读取下载源文件到目标文件中
while((len = bis.read(b)) != -1){
    //写入读取的数据
```

```
            bos.write(b,0,len);
        }
        //关闭缓冲输入流
        bis.close();
        //关闭输入流
        is.close();
        //关闭缓冲输出流
        bos.close();
        //关闭输出流对象
        os.close();
    }
}
```

2. DownLoadFileController 类的 Servlet 节点配置

```
<Servlet>
    <Servlet-name>downloadfile</Servlet-name>
    <Servlet-class>chapter11.DownLoadFileController</Servlet-class>
</Servlet>
```

3. DownLoadFileController 类的 Servlet 映射

```
<Servlet-mapping>
    <Servlet-name>downloadfile</Servlet-name>
    <url-pattern>/downloadfile</url-pattern>
</Servlet-mapping>
```

4. 下载页面 downloadfile.jsp

```
<%--
  Created by IntelliJ IDEA.
  User: wph-pc
  Date: 2018/12/4
  Time: 22:03
  To change this template use File | Settings | File Templates.
--%>
<%@ page contentType="text/html;charset=UTF-8" language="java"
isELIgnored="false" %>
<%@ include file="../header/init_bootstrap.jsp"%>
<html>
<head>
    <title>文件下载</title>

</head>
<body class="container">
<h2>第11章：文件下载案例</h2>
<hr>
<div class="panel panel-default">
    <div class="panel-heading panel-primary clearfix">
        <h3>文件下载技术</h3>
    </div>
    <div class="panel-body">
        <div class="input-group">
            <input type="text" class="form-control" id="txtFileName"
placeholder="请输入文件名称，支持模糊查询">
```

```html
                <span class="input-group-btn">
                    <button class="btn btn-default" type="button" id="btnFindFiles">查找</button>
                </span>
            </div>

            <table class="table table-responsive">
                <thead>
                <tr>
                    <th>序号</th>
                    <th>文件名称</th>
                    <th>文件类型</th>
                    <th>上传日期</th>
                    <th>大小</th>
                    <th>操作</th>
                </tr>
                </thead>
                <tbody id="tFiles">

                </tbody>
            </table>
        </div>
    </div>

    <script>
        function findFiles() {
            var con=new Object();
            con.fileName=$("#txtFileName").val();
            doData("/files",con,function (data) {
                if (data.obj!=undefined && data.obj!=null){
                    $("#tFiles").empty();
                    for(var i=0;i<data.obj.length;i++){
                        $("#tFiles").append("<tr><td>"+(i+1)+
                            "</td><td>"+data.obj[i].name+"</td><td>"+
                            data.obj[i].type+"</td><td>"+
                            formatDateTimebox(new Date(data.obj[i].createDate)) +
"</td><td>"+(data.obj[i].size/1024).toFixed(2)+
                            "<span style='color:deepskyblue'><strong>KB</strong></span></td><td>"+
                            "<a href='/jspweb/downloadfile?fileID="+data.obj[i].id+"' class='btn btn-xs'>下载</a></td></tr>");
                    }
                }
            },false);
        }
        $(function () {
            findFiles();
            $("#btnFindFiles").click(function () {
                findFiles();
            });
        });
```

```
        </script>
    </body>
</html>
```

运行结果

启动运行 downloadfile.jsp 页面，单击"下载"按钮，就可以实现文件下载。

习题

开发一个简单的个人相册管理系统，通过该系统，可以将个人的照片上传到服务器，也可以随时浏览或下载指定的照片，具体功能如下：

1．系统要求具备单点登录功能；
2．系统要求具备功能授权安全管理；
3．每个用户只能管理自己的相册，用户之间的项目不具有联系；
4．对照片可以进行分类管理，照片分类具有树形结构；
5．照片的主要属性包括：照片编号（自动编号，唯一性）、名称、日期、说明、存放地址、所属分类等基本信息；
6．照片信息要求存放在数据库中，照片附件存放在服务器上；能够通过照片"存放地址"属性查找到存放在服务器上的照片；
7．照片案例分类列表显示，支持按名称模糊查询功能；
8．照片支持多文件上传，并提供友好的上传进度支持，建议采用本章的文件上传技术；
9．能够实现照片下载，但是不能直接通过链接照片的服务器地址下载，需要使用本章的安全下载技术；
10．浏览照片不能直接通过链接照片服务器地址查看，需要转换处理。

反侵权盗版声明

电子工业出版社依法对本作品享有专有出版权。任何未经权利人书面许可，复制、销售或通过信息网络传播本作品的行为；歪曲、篡改、剽窃本作品的行为，均违反《中华人民共和国著作权法》，其行为人应承担相应的民事责任和行政责任，构成犯罪的，将被依法追究刑事责任。

为了维护市场秩序，保护权利人的合法权益，我社将依法查处和打击侵权盗版的单位和个人。欢迎社会各界人士积极举报侵权盗版行为，本社将奖励举报有功人员，并保证举报人的信息不被泄露。

举报电话：（010）88254396；（010）88258888

传　　真：（010）88254397

E-mail：　dbqq@phei.com.cn

通信地址：北京市万寿路173信箱
　　　　　电子工业出版社总编办公室

邮　　编：100036